T0253066

Applied AI and Humanoid Robotics for the Ultra-Smart Cyberspace

Eduard Babulak
National Science Foundation, USA

A volume in the Advances in
Computational Intelligence and
Robotics (ACIR) Book Series

Published in the United States of America by
    IGI Global
    Engineering Science Reference (an imprint of IGI Global)
    701 E. Chocolate Avenue
    Hershey PA, USA 17033
    Tel: 717-533-8845
    Fax: 717-533-8661
    E-mail: cust@igi-global.com
    Web site: http://www.igi-global.com

Library of Congress Cataloging-in-Publication Data

CIP DATA PROCESSING

Applied AI and Humanoid Robotics for the Ultra-Smart Cyberspace
 Eduard Babulak
  2024 Engineering Science Reference

ISBN: 9798369323991(hc) I ISBN: 9798369348840(sc) I eISBN: 9798369324004

This book is published in the IGI Global book series Advances in Computational Intelligence and Robotics (ACIR) (ISSN: 2327-0411; eISSN: 2327-042X)

British Cataloguing in Publication Data
A Cataloguing in Publication record for this book is available from the British Library.

All work contributed to this book is new, previously-unpublished material.
The views expressed in this book are those of the authors, but not necessarily of the publisher.

For electronic access to this publication, please contact: eresources@igi-global.com.

Advances in Computational Intelligence and Robotics (ACIR) Book Series

ISSN:2327-0411
EISSN:2327-042X

Editor-in-Chief: Ivan Giannoccaro, University of Salento, Italy

MISSION

While intelligence is traditionally a term applied to humans and human cognition, technology has progressed in such a way to allow for the development of intelligent systems able to simulate many human traits. With this new era of simulated and artificial intelligence, much research is needed in order to continue to advance the field and also to evaluate the ethical and societal concerns of the existence of artificial life and machine learning.

The **Advances in Computational Intelligence and Robotics (ACIR) Book Series** encourages scholarly discourse on all topics pertaining to evolutionary computing, artificial life, computational intelligence, machine learning, and robotics. ACIR presents the latest research being conducted on diverse topics in intelligence technologies with the goal of advancing knowledge and applications in this rapidly evolving field.

COVERAGE

- Brain Simulation
- Evolutionary Computing
- Computer Vision
- Computational Logic
- Robotics
- Fuzzy Systems
- Computational Intelligence
- Intelligent Control
- Artificial Life
- Algorithmic Learning

IGI Global is currently accepting manuscripts for publication within this series. To submit a proposal for a volume in this series, please contact our Acquisition Editors at Acquisitions@igi-global.com or visit: http://www.igi-global.com/publish/.

Titles in this Series

For a list of additional titles in this series, please visit:
http://www.igi-global.com/book-series/advances-computational-intelligence-robotics/73674

AI and IoT for Proactive Disaster Management
Mariyam Ouaissa (Chouaib Doukkali University, Morocco) Mariya Ouaissa (Cadi Ayyad University, Morocco) Zakaria Boulouard (Hassan II University, Casablanca, Morocco) Celestine Iwendi (University of Bolton, UK) and Moez Krichen (Al-Baha University, Saudi Arabia)
Engineering Science Reference • © 2024 • 299pp • H/C (ISBN: 9798369338964) • US $355.00

Utilizing AI and Machine Learning for Natural Disaster Management
D. Satishkumar (Nehru Institute of Technology, India) and M. Sivaraja (Nehru Institute of Technology, India)
Engineering Science Reference • © 2024 • 340pp • H/C (ISBN: 9798369333624) • US $315.00

Shaping the Future of Automation With Cloud-Enhanced Robotics
Rathishchandra Ramachandra Gatti (Sahyadri College of Engineering and Management, India) and Chandra Singh (Sahyadri College of Engineering and Management, India)
Engineering Science Reference • © 2024 • 431pp • H/C (ISBN: 9798369319147) • US $345.00

Bio-inspired Swarm Robotics and Control Algorithms, Mechanisms, and Strategies
Parijat Bhowmick (Indian Institute of Technology, Guwahati, India) Sima Das (Bengal College of Engineering and Technology, India) and Farshad Arvin (Durham University, UK)
Engineering Science Reference • © 2024 • 261pp • H/C (ISBN: 9798369312773) • US $315.00

For an entire list of titles in this series, please visit:
http://www.igi-global.com/book-series/advances-computational-intelligence-robotics/73674

701 East Chocolate Avenue, Hershey, PA 17033, USA
Tel: 717-533-8845 x100 • Fax: 717-533-8661
E-Mail: cust@igi-global.com • www.igi-global.com

I would like to dedicate this Book to my family and friends who provided me with their kind support and encouragement during the Book project.

Table of Contents

Foreword *by Jaydip Sen* xvii

Foreword *by Benjamin Fabian* xviii

Foreword *by Audrey Huong* xix

Foreword *by Giuseppe Carbone* xx

Preface xxi

Acknowledgment xxviii

Introduction xxix

Chapter 1
The Applied Artificial Intelligence Current State-of-the-Art and Greatest
Challenges: Applied AI and Humanoid Robotics..1
    Zahira Tabasssum, HKBK College of Engineering, India
    Hajira Siddiqua, HKBK College of Engineering, India
    Sufia Banu, HKBK College of Engineering, India
    Anees Fathima, HKBK College of Engineering, India
    Noor Ayesha, HKBK College of Engineering, India
    Rashmi Rani Samataray, HKBK College of Engineering, India

Chapter 2
Harnessing Applied AI: Transforming Industry and Business............................37
    Beulah Viji Christiana, Department of Master of Business
        Administration, Panimalar Engineering College, Chennai, India
    Krishnamohan Reddy Kunduru, Department of Engineering Design,
        Overhead Door Corporation, Lewisville, USA
    J. Chandrasekar, Department of Social Work, Madras School of Social
        Work, Chennai, India

B. Charith, Department of Management and Commerce, International
  Institute of Business Studies, Bengaluru, India
M. Thangatamilan, Department of Electronics and Instrumentation
  Engineering, Kongu Engineering College, Perundurai, India
A. Rajendra Prasad, Department of Mechanical Engineering, Sri Sai
  Ram Engineering College, Chennai, India

Chapter 3
AI and Blockchain Revolution in Communication Education: Future of Work
in the Age of Industrial Robotics 63
  C. V. Suresh Babu, Hindustan Institute of Technology and Science, India
  Sudhir Manoharan, Hindustan Institute of Technology and Science,
  India

Chapter 4
Elevating Performance for Enhancing AI-Powered Humanoid Robots
Through Innovation..85
  Krishnamohan Reddy Kunduru, Department of Engineering Design,
    Overhead Door Corporation, Lewisville, USA
  Yagya Dutta Dwivedi, Department of Aeronautical Engineering,
    Institute of Aeronautical Engineering, Hyderabad, India
  R. Aruna, Department of Electronics and Communication Engineering,
    AMC Engineering College, Bengaluru, India
  G. R. Thippeswamy, Department of Computer Science and Engineering,
    Don Bosco Institute of Technology, Bengaluru, India
  Subramanian Selvakumar, Department of Electrical and Computer
    Engineering, Bahir Dar Institute of Technology, Ethiopia
  M. Sudhakar, Department of Mechanical Engineering, Sri Sai Ram
    Engineering College, Chennai, India

Chapter 5
Revolutionizing Friction Stir Welding With AI-Integrated Humanoid Robots . 120
  B. Shamreen Ahamed, Department of Computer Science and
    Engineering, Sathyabama Institute of Science and Technology,
    Chennai, India
  Katragadda Sudhir Chakravarthy, Department of Mechanical
    Engineering, PACE Institute of Technology and Sciences, India
  Jeswin Arputhabalan, Department of Mechanical Engineering, Sri Sai
    Ram Institute of Technology, Chennai, India
  K. Sasirekha, Department of Computer Science and Business Systems,
    R.M.D. Engineering College, Kavaraipattei, India
  R. Malkiya Rasalin Prince, Department of Mechanical Engineering,

Karunya Institute of Technology and Sciences, Coimbatore, India
*S. Boopathi, Mechanical Engineering, Muthayammal Engineering
College, Namakkal, India*
*S. Muthuvel, Department of Mechanical Engineering, Kalasalingam
Academy of Research and Education, Srivilliputhur, India*

Chapter 6
Organized Ways to Increase the Fatigue of Mechanical Products Such as
Freezer Drawer Based on Quantum-Transferred Failure Model and Sample
Size...145
Seongwoo Woo, Reliability Association of Korea, North Korea
Yimer M. Hassen, Ethiopian Technical University, Ethiopia
Gezae Mebrahtu, Ethiopian Technical University, Ethiopia
Hadush Tedros Alem, Ethiopian Technical University, Ethiopia
Dennis L. O'Neal, Baylor University, USA

Chapter 7
Navigating the Future of Ultra-Smart Computing Cyberspace: Beyond
Boundaries 170
*N. Venkateswaran, Department of Master of Business Administration,
Panimalar Engineering College, Chennai, India*
*Krishnamohan Reddy Kunduru, Department of Engineering Design,
Overhead Door Corporation, Lewisville, USA*
*Nanda Ashwin, Department of Computer Science and Engineering
(IoT&CSBT), East Point College of Engineering and Technology,
Bangalore, India*
*C. S. Sundar Ganesh, Department of Electrical and Electronics
Engineering, Karpagam College of Engineering, Coimbatore, India*
*N. Hema, Department of Information Science and Engineering, RNS
Institute of Technology, Bangalore, India*
*Sampath Boopathi, Department of Mechanical Engineering,
Muthayammal Engineering College (Autonomous), Namakkal, India*

Chapter 8
Current and Future Research Directions...200
Himadri Sekhar Das, Haldia Institute of Technology, India

Conclusion 232

Compilation of References 239

Related References 256

About the Contributors 282

Index 285

Detailed Table of Contents

Foreword *by Jaydip Sen* xvii

Foreword *by Benjamin Fabian* xviii

Foreword *by Audrey Huong* xix

Foreword *by Giuseppe Carbone* xx

Preface xxi

Acknowledgment xxviii

Introduction xxix

Chapter 1
The Applied Artificial Intelligence Current State-of-the-Art and Greatest
Challenges: Applied AI and Humanoid Robotics 1
    Zahira Tabasssum, HKBK College of Engineering, India
    Hajira Siddiqua, HKBK College of Engineering, India
    Sufia Banu, HKBK College of Engineering, India
    Anees Fathima, HKBK College of Engineering, India
    Noor Ayesha, HKBK College of Engineering, India
    Rashmi Rani Samataray, HKBK College of Engineering, India

This chapter explores the intersection of applied artificial intelligence (AI) and humanoid robotics, focusing on their combined potential to revolutionize various aspects of human life. The integration of AI techniques within humanoid robots presents opportunities to enhance human-robot interaction, improve task efficiency, and advance technological capabilities. Through a comprehensive analysis of current research and practical applications, this chapter elucidates the synergy between applied AI and humanoid robotics, discussing its implications across diverse fields such as healthcare, education, manufacturing, and entertainment. Furthermore, it examines challenges and ethical considerations associated with this integration, emphasizing the importance of responsible development and deployment of AI-driven humanoid robots.

Chapter 2

Harnessing Applied AI: Transforming Industry and Business............................37

Beulah Viji Christiana, Department of Master of Business
Administration, Panimalar Engineering College, Chennai, India
Krishnamohan Reddy Kunduru, Department of Engineering Design,
Overhead Door Corporation, Lewisville, USA
J. Chandrasekar, Department of Social Work, Madras School of Social
Work, Chennai, India
B. Charith, Department of Management and Commerce, International
Institute of Business Studies, Bengaluru, India
M. Thangatamilan, Department of Electronics and Instrumentation
Engineering, Kongu Engineering College, Perundurai, India
A. Rajendra Prasad, Department of Mechanical Engineering, Sri Sai
Ram Engineering College, Chennai, India

This chapter delves into the transformative potential of applied artificial intelligence (AI) in various industries, highlighting its role in optimizing processes across sectors like manufacturing, logistics, and healthcare. It also discusses how AI is reshaping business operations, from sales and marketing to customer service and human resources, through the adoption of AI-powered tools and platforms. The chapter provides a comprehensive overview of AI's impact on modern business ecosystems. The chapter discusses the use of AI to enhance customer experiences, focusing on personalized interactions, tailored recommendations, and improved satisfaction across industries. It also addresses ethical considerations, governance frameworks, and future trends in AI. The chapter highlights the transformative potential of AI and advocates for continuous adaptation and innovation to fully utilize its benefits in industry and business.

Chapter 3

AI and Blockchain Revolution in Communication Education: Future of Work in the Age of Industrial Robotics 63

C. V. Suresh Babu, Hindustan Institute of Technology and Science, India
Sudhir Manoharan, Hindustan Institute of Technology and Science,
India

This chapter provides a comprehensive overview of AI and blockchain's role in communication education. It establishes the context, objectives, and significance of these technologies. The subsequent sections explore AI's impact in communication education, including content creation, personalization, and communication analysis.

It also delves into blockchain's contributions, focusing on trust and transparency. The chapter addresses employment patterns, discussing automation's impact and the need for adaptability. It examines implications for professionals and educators, emphasizing ethical considerations and curriculum integration. Additionally, it explores enhancing pedagogy with AI and blockchain, offers strategic recommendations, showcases case studies, and looks ahead, emphasizing ethics and human-centered approaches.

Chapter 4
Elevating Performance for Enhancing AI-Powered Humanoid Robots
Through Innovation...85

Krishnamohan Reddy Kunduru, Department of Engineering Design,
Overhead Door Corporation, Lewisville, USA
Yagya Dutta Dwivedi, Department of Aeronautical Engineering,
Institute of Aeronautical Engineering, Hyderabad, India
R. Aruna, Department of Electronics and Communication Engineering,
AMC Engineering College, Bengaluru, India
G. R. Thippeswamy, Department of Computer Science and Engineering,
Don Bosco Institute of Technology, Bengaluru, India
Subramanian Selvakumar, Department of Electrical and Computer
Engineering, Bahir Dar Institute of Technology, Ethiopia
M. Sudhakar, Department of Mechanical Engineering, Sri Sai Ram
Engineering College, Chennai, India

This chapter explores strategies for enhancing the performance of AI-powered humanoid robots through innovation. It begins with an overview of the current landscape of humanoid robots, highlighting their diverse applications across industries. The review examines existing performance metrics and identifies areas for improvement. Subsequent sections delve into specific avenues for innovation, including advancements in cognitive capabilities, motor skills, emotional intelligence, and human-robot interaction. Leveraging machine learning techniques for continuous improvement is also explored. Ethical considerations, such as privacy concerns and bias mitigation, are addressed, along with challenges associated with societal impact. The chapter concludes with case studies showcasing successful implementations of performance-enhancing strategies and outlines potential future directions for research and development in the field.

Chapter 5

Revolutionizing Friction Stir Welding With AI-Integrated Humanoid Robots . 120

B. Shamreen Ahamed, Department of Computer Science and Engineering, Sathyabama Institute of Science and Technology, Chennai, India

Katragadda Sudhir Chakravarthy, Department of Mechanical Engineering, PACE Institute of Technology and Sciences, India

Jeswin Arputhabalan, Department of Mechanical Engineering, Sri Sai Ram Institute of Technology, Chennai, India

K. Sasirekha, Department of Computer Science and Business Systems, R.M.D. Engineering College, Kavaraipattei, India

R. Malkiya Rasalin Prince, Department of Mechanical Engineering, Karunya Institute of Technology and Sciences, Coimbatore, India

S. Boopathi, Mechanical Engineering, Muthayammal Engineering College, Namakkal, India

S. Muthuvel, Department of Mechanical Engineering, Kalasalingam Academy of Research and Education, Srivilliputhur, India

This chapter explores the use of AI-integrated humanoid robots in friction stir welding (FSW), a crucial process for joining materials without melting. By combining AI capabilities with humanoid robots' dexterity and adaptability, significant advancements can be achieved. AI algorithms can improve precision and accuracy by continuously analyzing real-time sensor data, while AI-powered predictive maintenance can minimize downtime and enhance efficiency. AI-enabled robots in FSW increase automation, reduce human operator reliance, and minimize safety risks in hazardous environments. However, challenges such as cybersecurity concerns, regulatory hurdles, and ethical implications require careful consideration. Future research should focus on developing advanced AI algorithms, optimizing robot-human collaboration, and exploring new applications beyond traditional materials. The approach offers precision, efficiency, and safety, but necessitates interdisciplinary collaboration, strategic investment, and proactive addressing of technological, ethical, and regulatory challenges.

Chapter 6
Organized Ways to Increase the Fatigue of Mechanical Products Such as
Freezer Drawer Based on Quantum-Transferred Failure Model and Sample
Size...145

Seongwoo Woo, Reliability Association of Korea, North Korea
Yimer M. Hassen, Ethiopian Technical University, Ethiopia
Gezae Mebrahtu, Ethiopian Technical University, Ethiopia
Hadush Tedros Alem, Ethiopian Technical University, Ethiopia
Dennis L. O'Neal, Baylor University, USA

To improve the fatigue failure of systems, parametric accelerated life testing (ALT) is suggested. It includes (1) BX life scheme, (2) load evaluation, (3) a tailored representative of ALTs with adjustments, and (4) a calculation of whether product gets to the goal for the BX life. A quantum-transferred life-stress failure approach and sample size are recommended. As a case examination, the reliability of new drawer has been studied. In the 1st ALT, the handles fractured because of structural defects. As action plans, the whole handle width by providing an enhanced design that could correct the failures was enlarged. In the 2nd ALT, the slide rails of Freezer drawer also were being cracked and fractured because they did not have sufficient capacity to withstand the repeated food load in the Freezer drawer. To upgrade the design of slide rails in the drawer, additional strengthened ribs and boss, and an internal chamber in both rails were attached. After parametric ALTs, the altered Freezer drawer is anticipated to fulfill the lifetime aim – B1 life of 10 years.

Chapter 7

Navigating the Future of Ultra-Smart Computing Cyberspace: Beyond
Boundaries 170

N. Venkateswaran, Department of Master of Business Administration,
    Panimalar Engineering College, Chennai, India
Krishnamohan Reddy Kunduru, Department of Engineering Design,
    Overhead Door Corporation, Lewisville, USA
Nanda Ashwin, Department of Computer Science and Engineering
    (IoT&CSBT), East Point College of Engineering and Technology,
    Bangalore, India
C. S. Sundar Ganesh, Department of Electrical and Electronics
    Engineering, Karpagam College of Engineering, Coimbatore, India
N. Hema, Department of Information Science and Engineering, RNS
    Institute of Technology, Bangalore, India
Sampath Boopathi, Department of Mechanical Engineering,
    Muthayammal Engineering College (Autonomous), Namakkal, India

Ultra-smart computing cyberspace is a paradigm shift that combines artificial intelligence, augmented reality, and advanced networking technologies, transforming how we interact with digital environments. This integration offers unprecedented personalization, efficiency, and connectivity, blurring traditional computing boundaries and presenting challenges and opportunities in the ever-evolving technology landscape. Ultra-smart computing cyberspace presents opportunities for creativity, collaboration, and commerce, but also presents challenges such as privacy concerns, cybersecurity threats, and ethical considerations. To address these, industry stakeholders, policymakers, and technologists must establish robust frameworks to safeguard user rights and ensure responsible innovation. However, by leveraging data-driven insights and human-centered design principles, organizations can unlock transformative value and stay ahead in the competitive digital landscape.

Chapter 8

Current and Future Research Directions 200

Himadri Sekhar Das, Haldia Institute of Technology, India

The union of machine intelligence (AI) and manlike robotics has brought about extraordinary progresses in miscellaneous rules, promising the concoction of an extreme-smart information technology. These branches investigate current research trends and future guidance in used AI and manlike electronics for the growth of the extreme-smart cyberspace. The authors argue key sciences, challenges, and potential uses in various fields to a degree healthcare, education, amusement, and manufacturing. Additionally, they investigate moral concerns and societal impacts guide the unification of AI and manlike science into information technology. This comprehensive review aims to support acumens into the developing countryside of

AI and manlike robotics research, leading future endeavors towards achieving the thorough potential of the extreme-smart computer network.

Conclusion 232

Compilation of References 239

Related References 256

About the Contributors 282

Index 285

Foreword

We find ourselves in an era where the fusion of artificial intelligence (AI) and robotics reshapes our digital landscape. *Applied AI and Humanoid Robotics for the Ultra-Smart Cyberspace* offers a profound exploration into these dynamic domains. Within this volume lies a diverse tapestry of insights, delving into the myriad applications and ramifications of AI and humanoid robotics in our interconnected world. From healthcare to manufacturing, from education to entertainment, the pervasive influence of intelligent machines is palpable. This collection stands as a tribute to the visionaries, researchers, and engineers driving technological frontiers. Through its thought-provoking chapters, readers embark on a journey through the latest advancements, emerging trends, and ethical quandaries that define the future of AI and robotics. Despite the boundless promise of AI and robotics, they bring forth complex challenges. Matters of transparency, accountability, and ethics loom large as we navigate this era of innovation. Hence, this volume advocates for a critical examination, recognizing the intricate interplay between technology and society. At this juncture, as we stand on the threshold of a new epoch characterized by intelligent machines and ultra-smart cyberspaces, it falls upon us to chart a course guided by prudence and empathy. By fostering collaboration and interdisciplinary dialogue, we can harness the transformative potential of AI and robotics while upholding ethical principles and societal values. In conclusion, *Applied AI and Humanoid Robotics for the Ultra-Smart Cyberspace* emerges as a beacon of knowledge and inspiration, beckoning scholars, practitioners, and enthusiasts to explore and contribute to the unfolding narrative of AI and robotics.

Jaydip Sen
Praxis Business School, Kolkata, India

Foreword

With the current rapid advances in research, innovation, and development in the field of Artificial Intelligence (AI), Humanoid Robotics (HR) and Ultrasmart Cyberspace (USC), humanity is on the brink of a new era of technological revolution that will transform the way we live, communicate, learn, and work. The 24/7 access to information via ultra-fast, ubiquitous Internet will also provide a robotic communication platform, enabling robots to share their knowledge base and could exponentially increase their AI-driven performance. In parallel with the great potential of the transformation from Industry 4.0, 5.0 to 6.0, AI-driven humanoid robotics will have important impact on future smart health, education, industry, business, and government worldwide. This book contains a collection of selected contributions that highlight the latest developments in the field of applied AI in HR and USC, while also promoting the formation of global interdisciplinary research teams for the benefit of humankind.

Benjamin Fabian
Technical University of Applied Sciences Wildau, Germany

Foreword

Given the current fast advances in research, innovation and developments in the field of Artificial Intelligence (AI), Humanoid Robotics (HR) and the Ultra-Smart Cyberspace (USC), the humanity is yet in the new era of technological revolution that will transform the way we live, communicate, study and work. The 24/7 accessibility to information via ultra-fast ubiquitous Internet, apart from the public, is becoming a Robotic Communication platform, enabling Robots to share their Knowledge Base and exponentially increase their AI-driven intelligence. Parallel to the great potential transformation from Indutry4.0-5.0 to 6.0, AI-driven Humanoid Robotics will have an important impact on Future Smart Health, Education, industry, Business and Government worldwide. Professor Eduard's Book presents a collection of selected papers bringing to light the latest developments in the field of Applied AI in the field of HR and USC, while promoting the creation of global interdisciplinary research teams for the betterment of mankind. Thank you for producing this much anticipated and valuable Book!

Audrey Huong
Universiti Tun Hussein Onn Malaysia, Malaysia

Foreword

The ongoing advancements in Artificial Intelligence (AI), Humanoid Robotics (HR), and the Ultra-Smart Cyberspace (USC) are shaping a new era of technological evolution. This book presents a collection of selected papers highlighting the latest developments in Applied AI within the fields of HR and USC. Chapters delve into topics such as the current state-of-the-art and challenges in Applied Artificial Intelligence, the transformative impact of AI on industry and business, and the integration of AI and blockchain in communication and education. Other chapters explore innovations in enhancing the performance of AI-powered humanoid robots, revolutionizing manufacturing processes with AI-integrated robotics, and navigating the future of ultra-smart computing cyberspace. By bringing together insights from reputed researchers and practitioners, this book aims to foster interdisciplinary collaboration and drive forward the advancement of AI technologies. It provides valuable insights into current research directions and offers a glimpse into the future of AI-driven innovations in various domains.

Giuseppe Carbone
University of Calabria, Italy

Preface

With the rise of industrial revolution in the 19th century, the factory workers have been force to hard manual labor with very short time to rest and to recover during their working days. To get to work people had to walk before the invention of the transportation vehicles.

The meaning of the word WORK in Slovak language is *ROBOTA* and WORKING in Slovak language is *ROBIT*. To GET to WORK in Slovak languages is *IST do ROBOTY*, and finally SOMEONE WHO IS WORKING VERY HARD in Slovak language is *ROBIT ako ROBOT*.

Robot is drawn from an old Church Slavonic word, robota, for "servitude," "forced labor" or "drudgery." As a word, *robot* is a relative newcomer to the English language. It was the brainchild of a brilliant Czech playwright, novelist and journalist named Karel Čapek (1880-1938) who introduced it in his 1920 hit play, *R.U.R.*, or *Rossum's Universal Robots*.

Many dreamt about a day when they could get to work rested and clean, and most of their very hard manual labor would be replaced by machines and people/workers would be able monitor and control the manufacturing process on site or remotely.

Today, all this are almost forgotten memories of ancestors and our and younger generations today very seldom discuss or think about it. We all live in the age of the third millennium technological era where Industry 4.0 is transiting to 5.0., and future 6.0.

Today, we commute in beautiful cars, the most sophisticated fast trains, ships and the airplanes. Some, yet very few travel with the Space Shuttle to the Moon and watch our beautiful Blue Planet Mother Earth from highest altitudes while sharing their travel space with the Satellites far above in the hemisphere.

Perhaps, tomorrow our grandchildren will discuss the Human Technological Advancements with ultra-smart Humanoid Robots such as Sophia during their inter-galaxies journey, who will be able to project our history in three dimensional holographic - laser space and commentary facilitated by Super intelligent AI driven Ultra-Smart Cyberspace and Ultra Big Data Centre's.

I humbly believe that we are yet at the very beginning of new era of ultra-smart technological revolution that will transform the way we live, study, work and communicate. I sincerely hope that you will find the book educational, practical and inspiring while looking for ways to explore and master what may be considered today as impossible and futuristic.

- An overview of the subject matter

Given the current dynamic developments in the field of AI, Humanoid Robotics, Nano & Bio Technologies, New Materials, and Smart Medicine, with the ubiquitous access to high-speed Internet 24/7, the Ultra-smart Cyberspace is becoming reality.

The applications of AI with current and future dynamic trends in research, innovation and developments of Humanoid-Robotics, Computational Mechatronics, Smart Health, Cyber Security, and Ultra-Smart Cyber Technologies will contribute to betterment of mankind worldwide.

Given the current state of the art in the world of the Applied AI and Smart Computational Cyberspace, apart from technological advancements it is also very important to assess the social impact of the AI and Humanoid Robotics on society today and tomorrow.

- A description of where your topic fits in the world today

In support of future ultra-smart technological advancements it is important to have a better understanding of current and future AI driven Ultra-Smart Cyberspace, as well as value of the AI in the Industry 4.0 - 5.0 and future 6.0, future e-Services, Smart-Health, and other critical cyber infrastructures.

The areas of research in the field of AI, Humanoid Robotics, combined with the research in Ultra-Smart Computing are essential to humanity. In light of recent ongoing developments of Covid-19 crisis, having effective real-time application of Ultra-smart Cyberspace, driven by AI & Robotics and Big Data will support Next generation Smart-Health to save human lives.

The AI driven Smart Computational Systems are collecting, processing and analyzing a real-time medical data utilizing the Electronic Health Record (EHR) to fast treatment, prevention and healing of the wave of new viruses and diseases and ultimately safe human lives.

Due to Covid-19, the humanity lives in the most dramatic times, yet despite of its most negative impact it does also inspire dynamic innovation, research and developments in the world of AI, health, business, government, industry, plus others. The ranges and scale of possible applications of these Ultra-Smart Humanoid Robots has become ubiquitous and pervasive in societies of today and tomorrow worldwide.

- A description of the target audience

The book is primarily intended for the following type of audience.

1. Scholastic, Industry & Business Innovation and Development Research and Professional Communities worldwide: The book will be an essential reference for researchers in the subject domain having prerequisites of AI & Robotics.
2. University Faculty: The book presents practical quality technical know-how in the relevant field to the undergraduate/post-graduate students.
3. Postgraduates & Senior Undergraduates: The book will provide excellent resources to pursue higher studies in the form of research inspired by the different facets of AI & Robotics principles presented in book chapters with potential contributions to curriculum development.
4. Academic and corporate libraries: Adding this volume to libraries would enrich the content regarding the research base.
   - A description of the importance of each of the chapter submissions

The first chapter explores the intersection of applied artificial intelligence (AI) and humanoid robotics, focusing on their combined potential to revolutionize various aspects of human life. The integration of AI techniques within humanoid robots presents opportunities to enhance human-robot interaction, improve task efficiency, and advance technological capabilities.

Through a comprehensive analysis of current research and practical applications, this chapter elucidates the synergy between applied AI and humanoid robotics, discussing its implications across diverse fields such as healthcare, education, manufacturing, and entertainment. Furthermore, it examines challenges and ethical considerations associated with this integration, emphasizing the importance of responsible development and deployment of AI-driven humanoid robots.

The second chapter delves into the transformative potential of applied artificial intelligence (AI) in various industries, highlighting its role in optimizing processes across sectors like manufacturing, logistics, and healthcare. It also discusses how AI is reshaping business operations, from sales and marketing to customer service and human resources, through the adoption of AI-powered tools and platforms. The chapter provides a comprehensive overview of AI's impact on modern business ecosystems.

The chapter discusses the use of AI to enhance customer experiences, focusing on personalized interactions, tailored recommendations, and improved satisfaction across industries. It also addresses ethical considerations, governance frameworks, and future trends in AI. The chapter highlights the transformative potential of AI and advocates for continuous adaptation and innovation to fully utilize its benefits in industry and business.

The third chapter provides a comprehensive overview of AI and Blockchain's role in Communication Education. It establishes the context, objectives, and significance of these technologies. The subsequent sections explore AI's impact in Communication Education, including content creation, personalization, and communication analysis. It also delves into Blockchain's contributions, focusing on trust and transparency.

The chapter addresses employment patterns, discussing automation's impact and the need for adaptability. It examines implications for professionals and educators, emphasizing ethical considerations and curriculum integration. Additionally, it explores enhancing pedagogy with AI and Blockchain, offers strategic recommendations, showcases case studies, and looks ahead, emphasizing ethics and human-centered approaches.

The fourth chapter explores strategies for enhancing the performance of AI-powered humanoid robots through innovation. It begins with an overview of the current landscape of humanoid robots, highlighting their diverse applications across industries. The review examines existing performance metrics and identifies areas for improvement. Subsequent sections delve into specific avenues for innovation, including advancements in cognitive capabilities, motor skills, emotional intelligence, and human-robot interaction.

Leveraging machine learning techniques for continuous improvement is also explored. Ethical considerations, such as privacy concerns and bias mitigation, are addressed, along with challenges associated with societal impact. The chapter concludes with case studies showcasing successful implementations of performance-enhancing strategies and outlines potential future directions for research and development in the field.

The fifth chapter explores the use of AI-integrated humanoid robots in friction stir welding (FSW), a crucial process for joining materials without melting. By combining AI capabilities with humanoid robots' dexterity and adaptability, significant advancements can be achieved.

AI algorithms can improve precision and accuracy by continuously analyzing real-time sensor data, while AI-powered predictive maintenance can minimize downtime and enhance efficiency. AI-enabled robots in FSW increase automation, reduce human operator reliance, and minimize safety risks in hazardous environments. However, challenges such as cybersecurity concerns, regulatory hurdles, and ethical implications require careful consideration.

Future research should focus on developing advanced AI algorithms, optimizing robot-human collaboration, and exploring new applications beyond traditional materials. The approach offers precision, efficiency, and safety, but necessitates interdisciplinary collaboration, strategic investment, and proactive addressing of technological, ethical, and regulatory challenges.

The sixth chapter presents methods to improve the fatigue failure of systems operated by machine such as car, airplane, refrigerator, etc., parametric accelerated life testing (ALT) as a methodology of reliability tests that can find out the design defects is suggested.

The process includes: (1) parametric ALT used to establish the BX life, (2) load evaluation for designing an accelerated life test(s), (3) a tailored representative of ALTs with adjustments, and (4) a calculation of whether product gets to the goal for the BX life. A quantum-transferred life-stress failure approach, and sample size are recommended. As a case examination, the reliability of new freezer drawer system has been studied.

The seventh chapter discusses the Ultra-smart computing cyberspace as a paradigm shift that combines artificial intelligence, augmented reality, and advanced networking technologies, transforming how we interact with digital environments.

This integration offers unprecedented personalization, efficiency, and connectivity, blurring traditional computing boundaries and presenting challenges and opportunities in the ever-evolving technology landscape.

Ultra-smart computing cyberspace presents opportunities for creativity, collaboration, and commerce, but also presents challenges such as privacy concerns, cybersecurity threats, and ethical considerations. To address these, industry stakeholders, policymakers, and technologists must establish robust frameworks to safeguard user rights and ensure responsible innovation. However, by leveraging data-driven insights and human-centered design principles, organizations can unlock transformative value and stay ahead in the competitive digital landscape.

The eighth chapter presents the union of machine intelligence (AI) and manlike robotics has surpassed to extraordinary progresses in miscellaneous rules, promising the concoction of an extreme-smart information technology. These branches investigate current research trends and future guidance in used AI and manlike electronics for the growth of the extreme-smart cyberspace.

The author(s) argue key sciences, challenges, and potential uses in various fields to a degree healthcare, education, amusement, and manufacturing. Additionally, the authors investigate moral concerns and societal impacts guide the unification of AI and manlike science into information technology.

This comprehensive review aims to support acumens into the developing countryside of AI and manlike robotics research, leading future endeavors towards achieving the thorough potential of the extreme-smart computer network.

The tenth chapter presents conclusion and further research directions. The author present couple of philosophical questions that may have remained to be unanswered, such as:

1) Would any mother or anyone trust a Humanoid Robot with their new born baby?

2) Would anyone trust the AI to be in charge of their personal safety, privacy, economy, health and education?

3) Would the future Ultra-Smart Humanoid Robots be able to embrace/assimilate the Human Nature, Human values and to reason like us people?

4) Would future Ultra-Smart Humanoid Robot be able to embrace/assimilate Spirituality?

5) Plus they may be many more unanswered questions that will inspire scholars, industry and business practitioners to seek that new technological solutions that will bring Machines closer to Humanity and yet bring the betterment for all mankind.

   ○   A conclusion

The most recent advances in the Applied Artificial Intelligence, Smart Computation, Nano-Biotechnology, Mechatronics, etc. contribute to creating a future Ultra-Smart Human-like Intelligent Machines, which may be capable of assimilating a Human nature in its physical and emotional dimensions. These future Ultra-Smart Intelligent Machines will be far beyond sophistication of a current Sophia Smart Robot. There are currently ongoing discussions concerning the potential value and impact of the AI and Humanoid Robotics on society today and tomorrow.

Applied AI and Humanoid Robotics has become an integrated part of curriculum in the Academia, and there are many applications in Government, Business and Industry worldwide. The current dynamic advancements if the field of AI and Humanoid Robotics will continue impacting all aspect of human lives for the years to come.

The book main objective is to promote creation of *Global Center of Excellence in Applied AI & Humanoid Robotics* for research, innovation and development of the AI driven Ultra-Smart Computational Devices.

The second objectives is to promote a better understanding and clarity of the value that the AI and Humanoid Robotics present today and will present tomorrow to the world of the Ultra Smart Computational Cyberspace and ultimately to the society worldwide.

The third objective is to establish the Assessment Metrics for the AI & Humanoid Robotics Social Impact on Public and Scientific Communities worldwide.

The Book discusses the fundamental principles of Ethics and Professionalism, while providing assessment of the applied AI applications social impact on national and global levels.

It also promotes seamless creation of multidisciplinary teams of experts in the nation and worldwide. Reader may seek the answers to a question like: Will the AI be well understood and become part of our daily live or else?

The book brings clarity and better understanding of the value that AI and Humanoid Robotics brings to society, which will ultimately open gate to new opportunities for future innovation, research and development in this field of studies.

Eduard Babulak
National Science Foundation, USA

Acknowledgment

I would like to use this opportunity to express my sincere gratitude to contributors, reviewers, to Ms Eddinger, Ms Wagner, Ms Barrantes, Ms Travers, Ms McLouglin, Ms Wertman and the IGI Global Colleagues and Friends for their excellent support, kind guidance and good counsel during the course of this and my earlier projects. I would like to say great thank you to my family and friends for their kind support and encouragement during the Book project.

Editorial Advisory Board

Giuseppe Carbone, *DIMEG, University of Calabria, Italy*

Audrey Huong Kah Ching, *Universiti Tun Hussein Onn Malaysia, Malaysia*

S. N. Kumar, *Department of EEE, Amal Jyothi College of Engineering, Kerala, India*

Jaydip Sen, *Praxis Business School, Kolkata, India*

Introduction

RISE OF ROBOTS

According to Kevin M. Lynch and Frank C. Park, robotics was inspired by the creation of machines that can behave and think like humans. The attempt to create intelligent machines naturally examines why our human bodies are designed the way they are, how our limbs are coordinated, and how we learn and perform complex tasks. The sense that the fundamental questions in robotics are ultimately questions about us is part of what makes robotics such a fascinating and engaging endeavor (Lynch & Park, 2017).

Figure 1. Two-legged robot mimics human balance while running and jumping (MIT)

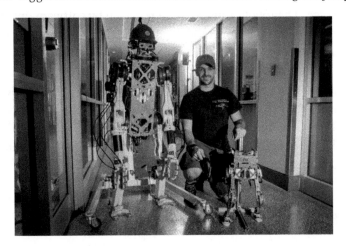

The tendency to think about robots as having a human-like appearance may stem from the origins of the term "robot." The word "robot" came into the popular consciousness on January 25, 1921, in Prague, with the first performance of Karel Capek's play, R.U.R. (Rossum's Universal Robots).

In R.U.R., an unseen inventor, Rossum, has created a race of workers that were made from a vat of biological parts, smart enough to replace a human in any job (therefore "universal"). Capek called the workers robots, a term derived from the Czech and Slovak word "robota," which is loosely translated as menial laborer. The term robot workers implied that the artificial creatures were strictly meant to be servants to free "real" people from any type of labor (Lynch & Park, 2017).

The robot consists of a sensor, controlling device, physical device, manipulator, and a programming testing device. The construction of a Robot brings together experts from mechanical engineering, electrical engineering, mathematics, and bio-chemistry. The intelligent robot should have following characteristics (Dubey et al., 2021):

- Sensing: to be able to recognize the surroundings and respond accordingly. The robots may not behave in all the environments in the same manner. Combining sensitivities to light (vision), touch, pressure, chemicals (smell), sound (hearing) and taste are important to enable the robot to function and work correctly in the particular environment.
- Movement: Mastery of Geometry is essential for robot to be capable to identify its surroundings – environment, and to operate correctly in given environment.
- Energy: Capability of robot to recognize its power source (i.e. battery) and to be charged are critical.
- Intelligence: The robotics intelligence is driven by software. Intelligent robot is capable to perform the operations set by the software. Perhaps, the futuristic robots may become smarter than us humans.

First industrial robot was developed the first industrial robot by George Devol, American inventor and founder of the first robotics company Unimation. The first industrial robot was a hydraulic arm called Unimate and served to lift the heavy objects. At the very beginning robots were considered to be a mechanical devices operated by hydralic mechanisms.

RISE OF SMART CYBERSPACE

Before Cyberspace, there was a simple idea presented by DARPA project of creating to connect to computational devices (i.e., personal computers), via a communication channel. In 1968, during my elementary schooling the idea of Internet was borne by the U.S. Advanced Research Projects Agency Network (ARPANET). It was a public

Figure 2. Arpanet (ARPANET)

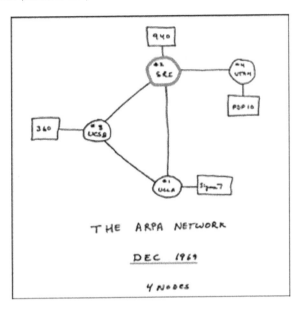

packet-switched computer network, first used in 1969 and finally decommissioned in 1989.

In today's world the diagram shown in Figure 2, may appear to be very simple, yet to engineering who work on the ARPANET, this was a great challenge and very complex engineering task.

Figure 3. Growth of ARPANET (ARPANET)

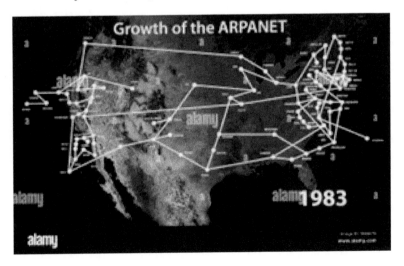

The ARPANET's main was created to enable communication among the academic and research community for research, innovation and development of new technologies. Figure 3 shows how the ARPANET developed in 1983.

Since, 1983 we have seen ARPANET transformation in today's Global Internet, which has become for us a 24/7 communication platform accessed from anywhere by anyone wishing to create and post the text, image, and video in his or her social community worldwide. Internet today is driven by Large Data Centers, like Facebook, Google, Microsoft, Samsung, Apple, IBM, Amazon, Baidu, plus other worldwide.

My generation remembers well a simple Analogue Rotary Telephone shown in Figure 4, operated by Electromagnetic Relays and Rotary Switches. Often, during a long distance call, users might have overheard the conversations by other users caused by the signal interference among the telephone lines.

Figure 4. Rotary Telephone (Google images)

My first job was to operate a PP51 Telephone Exchange Station shown in Figure 5, which produced lot of heat and noise during its operation, and was larger than a classroom or school gym.

Figure 5. Relay based Telephone Exchange

Figure 6. Black and White TV
Source: Google images

In early 70's, having a black and White TV and to have 2 TV channels was to many a luxury. Figure 6, shows a B&W TV, which as quite large and heavy to lift. Soon after my college graduation in 1976, I was working a TV Repairman which was quite challenging due to heat produced my Electro9nic Tubes and potential risk of electric shock.

The appreciation of history of very first Electronic Communication Technologies is essential to be able to imagine possible future Ultra-Smart Communication Technologies. The third millennium opened a new era of ultra-fast ubiquitous Internet and smart computing technologies, which created a platform for initiating a next level applied research in the next generation Ultra-Smart Computational Devices and Fully Automated Cyberspace. Given the current dynamic developments in the field of AI & Robotics, Big Data, Massive Data Storage and Ubiquitous access to high-speed Internet 24/7 for anyone worldwide, the term Smart Cyberspace is becoming well accepted reality.

Figure 7. Smart Cyberspace (source: Google)

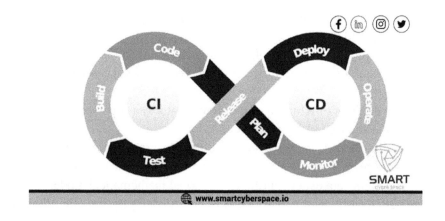

Apart from the Internet used by people, there is a new Internet used by robots, called a Robotic Internet. Today's sophisticated smart robots are capable of communicating and sharing knowledge base among themselves in support of augment learning. In other words, Machines are learning from Machines (MLM) and are driven by AI in ubiquitous and pervasive manner. Figure 7, shows a simplistic view of Smart Cyberspace as an open platform to Plan, Monitor, Operate, Deploy, Release, Test, Build, and Code.

RISE OF AI

The term "artificial intelligence" (AI) was first mentioned in 1856 by John McCarthy. Artificial intelligence (AI) is a science of making machines act intelligently. Some define AI as "the study of ideas that enable computers to be intelligent", or more specifically, "AI is the attempt to get the computer to do things that, for the moment, people are better at" (Lynch & Park, 2017).

Some refer to AI as intelligence algorithms that make machines capable of assimilating a human behavior and intelligence and to reason and act like humans (Dubey et al., 2021). The name artificial is used because the intelligence has been programmed into the machine which is different from the natural intelligence common to people.

AI is one of the disruptive technologies with powerful features and can be used in a wide variety of applications; for example, AI can be used to play games, monitor the health of patients and used as a traffic controller system (Dubey et al., 2021). The rise of the AI era triggered a fundamental change in the landscape ultra-smart computation applied in government, academia, industry, and business of modern enterprises worldwide. The AI's dynamic advancements open a new era of innovation, research and development of next generation ultra-smart computational technologies, which ultimately promote new opportunities and challenges for any enterprise striving to be competitive and relevant in the age of digital transformation and continually evolving geo-political climate.

Most common sub-branches of the AI include, Machine Learning, Speech recognition, Natural Language Processing, Planning, Expert Systems, Vision and Robotics (Fig.8).

The Venn diagram of AI in Figure 9, shows that it is a superset of AI, Machine Learning (ML) and Deep Learning (DL).

ML describes the learning behavior of machines programmed by software and data input. Machine are learning and make predictions based on statistical inference and identifying certain data patterns utilizing advanced mathematical models (Swedha & Dubey, 2018). ML works as backend of the AI defining specific algorithmic

Figure 8. The AI Sub-branches

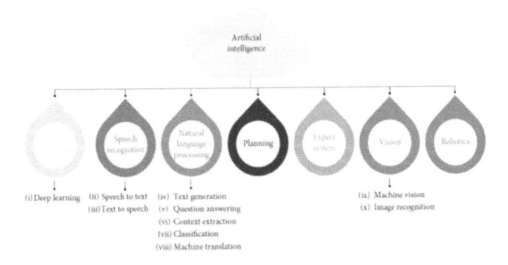

Figure 9. Venn diagram of AI, ML and DL

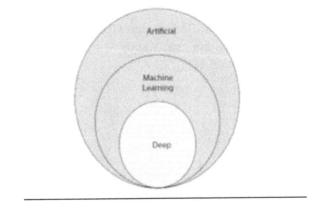

procedures that drive machine intelligence and enable accurate predictions (Kittur & Pais, 2019). ML is divided into four types:

- Supervised learning
- Unsupervised learning
- Semi-supervised learning
- Reinforcement learning

Figure 10. Representation of an Agent

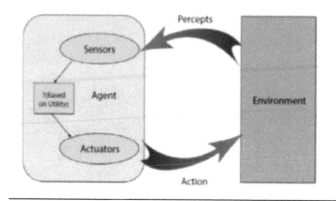

Agent: Refers to something that perceives the environment through sensors and performs action via actuators based on some predefined rules for which the agent is trained. Just like in humans, there are five sense (i.e., sight, sound, smell, taste, and touch), and based on the information of our senses we perform actions through our limbs, etc. A simple diagrammatic representation of an agent is shown in Figure 10 below.

The agents act as a backbone for AI techniques that govern how they are working and what sorts of applications they are dealing with. Based on utility there are different types of agents (Gayatri et al., 2018):

- Simple reflex agent
- Model-based agent
- Goal-based agent
- Utility-based agent

Deep learning is a method in artificial intelligence (AI) that teaches computers to process data in a way that is inspired by the human brain. Deep learning models can recognize complex patterns in pictures, text, sounds, and other data to produce accurate insights and predictions. Figure 11 shows DL vs ML. Given the same input, ML process the feature extractions, followed by classification leading to output, while DL process feature extraction and classification together leading to output.

Figure 11. Deep Learning vs Machine Learning

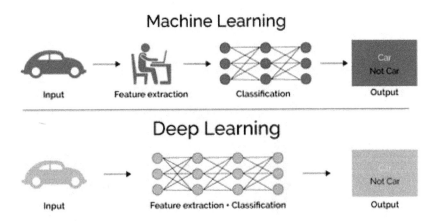

RISE OF HUMANOID ROBOTICS

The intelligent machines like Humanoid Robots like Sophia shown in Figure 12, have not yet reach the level of reasoning that is based of human fundamental values such as common sense and values (Babulak, 2023). However, values of humanities such as ethics, professionalism, and care are common but not always well understood by some people.

Future philosophical horizons may bring the AI driven intelligent computational machines like Humanoid Robots much closer to humans. Time may come when robots like Sophia may be accepted without any suspicion that they may cause any possible harm or danger to humans.

Development of Sophia brought together multidisciplinary global team of expert from number of disciplines, including Bio-Mechanical, Computer Engineering & Science, Social Sciences and others. Sophia is a result of long lasting intensive research, innovation and development working side-by-side with military, academia, industry, and business worldwide. Yet, Sophia is just a beginning of new era of future Ultra-Smart Humanoid Robots that will look like, act like and reason like us Humans.

One of the most challenging and somewhat philosophical questions for any researcher, expert, or engineer who may be working on developing a new generation of Humanoid Robots is to define a level of trust, reliability and sensitivity that these intelligent machines may reach in comparison to humans. One of the examples that my illustrate the current level of trust, reliability and sensitivity of intelligent machine is taking care of babies and or people who may some level of physical or mental deficiencies.

Figure 12. Sophia Smart Humanoid Robot (Google images)

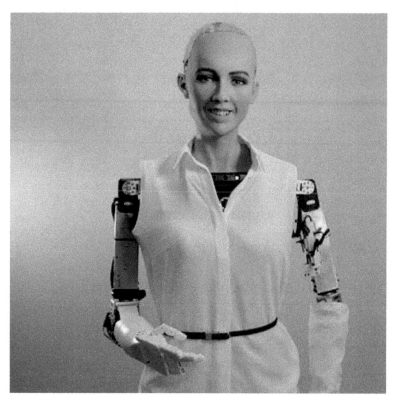

Perhaps most challenging would be to imagine a possible scenario in which a sophisticated Humanoid Robot today or tomorrow may be asked to take care of new born baby.

This scenario may require lot of imagination and experimentation while having a well-established set of metrics that would assess the Humanoid Robot human like capabilities that may warrant a certain level of trust that is required by a natural mother or person taking care of new born baby.

The metrics common parameters may include a human-like values, emotions, feelings, touch, perceptions, sensitivity, decision making, finesse, adaptation and care. Human taking care of baby may have a certain stage of maturity, medical condition, a daily routine. Human have breakfast, lunch, dinner, refreshment (i.e., tea, coffee, etc.)

Compare to human medical condition, a Humanoid Robot would have an External or Internal Power Supply, System Hardware and Software that may be prone to malfunction triggered by its design complexity and interconnectivity within its internal computational system and to control or monitoring remote systems connected by robotic communication network infrastructures and Internet (Ammanath, 2022, Nam & Lyons, 2020).

Given the required service of perfection 24/7, humanoid robot must and should not harm nor put in any danger a small baby due to its system imperfections and possible malfunction. Similar, any human taking care of small baby must and should provide perfect care 24/7.

THE SOCIAL IMPACT OF THE ARTIFICIAL INTELLIGENCE AND HUMANOID ROBOTICS

In resume, the AI and Humanoid Robotics today are at the very early stage of its evolution. However, the current results demonstrated by academia, industry and business show great potential of future applied research, innovation, development in the field of Human-like Intelligent Machines that may be well accepted in our daily lives as trusted and reliable companions .

Today, we humans may have opinion about the today's and future Humanoid Robots possible imperfections and lack of humanity. Perhaps the day may come when the Humanoid Robots may have opinion about our human imperfections and lack of machine-like reasoning (Pak, 2019).

Will humanity be able to accept the Machines that may be challenging their own intelligence and capabilities to reason? Perhaps time will show how far the Machines will be capable to augment their intelligence and capabilities, and most of all to learn how to work with people together for the betterment of mankind. Figure 13, shows an example of robot caring for an elderly person as a house companion.

Figure 13. Robot a House Companion (Aretove Technologies)

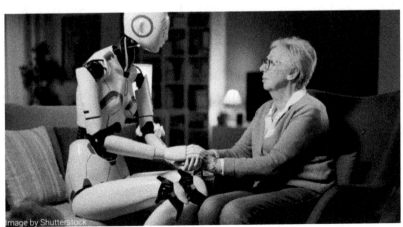

Figure 14. Humanity vs Machines (Cyber Industry)

To make sure that Future Ultra-Smart Humanoid Robots and Ultra-Smart Cyberspace will not take control of Human lives, it is essential to bring the researchers, innovators, developers and decision makes to come together a round table shown in Figure 14, to discuss, to brainstorm, and to plan future directions to make sure that Machines will not abuse, nor endanger in any way Human lives, but will be used and driven by idea to contribute to the betterment of mankind.

I humbly believe that working and learning from each other the Humanity will be able to master any challenges that future Ultra-Smart Technologies may bring tomorrow. Similar to Machines, Humanity willing to learn will be able to make amazing progress and their children and grandchildren will do the same, yet in very different world that many of us may not be able to imagine today.

RISE OF ULTRA-SMART FULLY AUTOMATED CYBERSPACE

The current advancements in Humanoid Robotics and Robotic Internet, Big Data, AI and Machine Learning, Tele-Medicine, in conjunction with collecting real-time data from the Electronic Health Record (EHR) in the nation and worldwide, as well as collections of antibodies contributes well to community worldwide aspirations to safe human lives and to restart the economies worldwide.

The areas of research in the field of robotics that are closely related to the modeling, motion generation, and control of humanoid robots continue to evolve rapidly worldwide. Research results in the fields of physics-based animation of articulated figures and the biomechanics of human movement are shown to share a number of common points. In light of currently ongoing developments of Covid-19 crisis, having effective real-time application of Artificial Intelligence & Robotics with the Big Data remotely control via Internet is essential.

These are most dramatic times for mankind worldwide, and yet despite of its most negative impact it does also inspire dynamic innovation, research and developments in the world of health, business, government, industry, plus., while promoting seamless creation of multidisciplinary teams of experts in the nation and worldwide.

The project promotes better understanding of the current and future dynamic trends in research, innovation and developments of cutting-edge technologies, Humanoid Robotics, AI, and smart cyber systems that may contribute effectively to people saving lives, and decision makers in the nation and worldwide.

The Sophia Humanoid Robot have opened a new chapter in Robotics' Evolution, and new beginning of bringing Artificial Intelligence to a level of Human-like Intelligence while exploring new ways of imagining, creating and implementing Humanoid Robots that may have computational, mechatronics, and interaction capabilities far beyond that of Sophia and today's most sophisticated so-called "AI driven Smart Humanoid Robots".

REFERENCES

Ammanath, B. (2022). *Trustworthy AI: A Business Guide for Navigating Trust and Ethics in AI.* Wiley.

Babulak, E. (2023). *The Third Milennium AI driven Humanoid Robots.* Swiss Cognitive https://swisscognitive.ch/2023/01/19/the-third-millennium-ai-driven-humanoid-robots/

Dubey, A. K., Abhishek Kumar, S., & Rakesh Kumar, N. Gayathri, & Das, P. (2021). AI and IoT-Based Intelligent Automation in Robotics. Wiley.

Gayatri, P., Venunath, M., Subhashini, V., & Umar, S. (2018). Securities and threats of Cloud Computing and Solutions. In *2018 2nd Int. Conf. Inven. Syst. Control, no. Icisc*, (pp. 1162–1166).

Kittur, A. S., & Pais, A. R. (2019). *A new batch verification scheme for ECDSA A˜ signatures.* Indian Academy of Sciences., doi:10.1007/s12046-019-1142-9

Lynch, K. M., & Park, F. C. (2017). *Modern Robotics: Mechanics, Planning, and Control*. Cambridge University Press.

Nam, C. S., & Lyons, J. B. (2020). *Trust in Human-Robot Interaction*. Elsevier.

Pak, R. (2019). *Living with Robots: Emerging Issues on the Psychological and Social Implications of Robotics*. Elsevier.

Robin, R. (2019). *Murphy, Introduction to AI Robotics*. Massachusetts Institute of Technology.

Swedha, K., & Dubey, T. (2018). Analysis of Web Authentication methods using Amazon Web Services. In *2018 9th Int. Conf. Comput. Commun. Netw. Technol.* (pp. 1–6).

Chapter 1
The Applied Artificial Intelligence Current State-of-the-Art and Greatest Challenges:
Applied AI and Humanoid Robotics

Zahira Tabasssum
HKBK College of Engineering, India

Anees Fathima
HKBK College of Engineering, India

Hajira Siddiqua
HKBK College of Engineering, India

Noor Ayesha
HKBK College of Engineering, India

Sufia Banu
HKBK College of Engineering, India

Rashmi Rani Samataray
HKBK College of Engineering, India

ABSTRACT

This chapter explores the intersection of applied artificial intelligence (AI) and humanoid robotics, focusing on their combined potential to revolutionize various aspects of human life. The integration of AI techniques within humanoid robots presents opportunities to enhance human-robot interaction, improve task efficiency, and advance technological capabilities. Through a comprehensive analysis of current research and practical applications, this chapter elucidates the synergy between applied AI and humanoid robotics, discussing its implications across diverse fields such as healthcare, education, manufacturing, and entertainment. Furthermore, it examines challenges and ethical considerations associated with this integration, emphasizing the importance of responsible development and deployment of AI-driven humanoid robots.

DOI: 10.4018/979-8-3693-2399-1.ch001

INTRODUCTION

Artificial Intelligence (AI) represents a revolutionary domain within computer science, aiming to replicate human-like intelligence in machines. Through advanced algorithms and computing power, AI systems can understand natural language, recognize patterns, and make decisions, among other tasks traditionally associated with human cognition (Y.Tong, 2024). In parallel, humanoid robots have emerged as a tangible application of AI, created to varied degrees to resemble the appearance and behaviour of humans. These robots integrate sophisticated sensory systems, processing capabilities, and mechanical actuators to interact with their environment and humans more naturally. Over the past few years, artificial intelligence (AI) interaction with humanoid robotics has attracted a lot of interest due to its potential to transform numerous domains, covering everything from entertainment to health. Humanoid robots, designed to resemble and interact with humans, offer unique advantages in facilitating natural communication and performing tasks in human-centric environments. By harnessing the power of applied AI methods including computer vision, natural language processing, and machine learning, these robots can exhibit intelligent behavior and adaptability, thereby enhancing their utility and effectiveness. This chapter provides a thorough rundown of how applied AI is shaping the capabilities and applications of humanoid robotics, revolutionizing the way humans interact with and benefit from robotic systems.

Figure 1. Humanoid robotics overview

Foundations of Applied AI in Humanoid Robotics

This section delves into the fundamental principles and technologies underpinning the integration of applied AI with humanoid robotics. It explores key concepts such as deep learning, reinforcement learning, and cognitive architectures, elucidating how these AI techniques enable robots to perceive, reason, and act in dynamic environments. Moreover, it discusses the role of sensor fusion and multimodal perception in enhancing the sensory capabilities of humanoid robots, enabling them to interpret and respond to human cues effectively.

Human-Robot Interaction (HRI)

Human-robot interaction (HRI) lies at the heart of applied AI in humanoid robotics, emphasizing seamless communication and collaboration between humans and robots. This section examines advancements in HRI techniques, including gesture recognition, emotion detection, and dialogue management, which enable robots to engage with users in natural and intuitive ways. Additionally, it explores the design principles for creating socially aware and empathetic robots capable of understanding and responding to human emotions and social norms.

Applications in Healthcare

Humanoid robots equipped with applied AI hold immense potential to revolutionize healthcare delivery by assisting medical professionals, supporting patients, and enhancing rehabilitation processes. This section explores how AI-driven humanoid robots are being deployed in hospitals, clinics, and care facilities to perform tasks such as patient monitoring, medication management, and physical therapy. Moreover, it discusses the role of telepresence robots in enabling remote healthcare consultations and improving access to medical expertise in underserved areas.

Transforming Education and Training

In the field of tutoring, AI-powered humanoid robots are redefining traditional teaching methodologies and personalized learning experiences. This section investigates how robots equipped with adaptive learning algorithms and educational content can engage students, provide personalized feedback, and facilitate interactive learning activities. Furthermore, it explores the use of humanoid robots as educational assistants and tutors, supporting students with special needs and promoting inclusive learning environments.

Advancements in Manufacturing and Industry

Applied AI in humanoid robotics is driving innovation in industrial and manufacturing environments, where robots play a crucial role in automating repetitive tasks, enhancing productivity, and ensuring workplace safety. This section examines how AI-powered humanoid robots are being integrated into assembly lines, warehouses, and logistics operations to perform complex manipulation tasks, quality inspections, and collaborative manufacturing processes. Additionally, it discusses the potential of cobots (collaborative robots) to work alongside human workers, optimizing production efficiency and flexibility.

Ethical Considerations and Future Directions

As AI-driven humanoid robotics continue to proliferate, it is imperative to address ethical concerns related to privacy, autonomy, and societal impact. This section explores ethical considerations surrounding the design, deployment, and regulation of AI-powered humanoid robots, emphasizing the need for transparency, accountability, and ethical guidelines. Moreover, it speculates on future trends and challenges in the field, highlighting opportunities for interdisciplinary research and collaboration to maximize the beneficial impact of applied AI in humanoid robotics while mitigating potential risks.

Related Works

Table 1. Existing approaches of applied AI in humanoid robotics

Author	Title	Methodology	Limitations
Demetris Vrontis (2021)	Artificial intelligence, robotics, advanced technologies and human resource management: a systematic review	To systematise the existing body of scholarly work on intelligent automation and identify the key areas of contribution and areas of difficulty for HRM.	Technology has a far more complicated role in HRM than just enhancing or changing HR procedures.
Okagbue (2023}	An in-depth analysis of humanoid robotics in higher education system	In order to urge African policymakers to recognise the importance of advancing the use of HR in African education, the paper will highlight the top ten most referenced authors and significant journals in the field of human resources.	Curriculum developers and educational policymakers in Africa should work to establish measures that will help African higher education institutions institutionalize humanoid robotics into their curricula.
Venkataswamy (2024)	Realization of Humanoid Doctor and Real-Time Diagnostics of Disease Using Internet of Things, Edge Impulse Platform, and ChatGPT.	A cloud artificial intelligence platform is used to train the historical patient data, and edge and medical IoT devices are used to obtain patient sample data for testing the model.	The absence of a multifeatured accurate model, accessibility, availability, and standardisation has led to a lack of faith in the humanoid doctor.
Pérez, L(2019)	Symbiotic human–robot collaborative approach for increased productivity and enhanced safety in the aerospace manufacturing industry	A cutting-edge advanced manufacturing technique applied in the aerospace industry that blends human adaptability with robot strength and consistency. Control, security, and interface elements are included in the suggested system architecture for the new collaborative manufacturing process.	There are numerous fields with potential uses, including other robot-reluctant sectors.

Role of AI in Humanoid Robotics

Artificial Intelligence's (AI) function in humanoid robotics is pivotal, driving advancements that blur the lines between human and machine interaction. AI serves as the cognitive backbone of humanoid robots, endowing them with the ability to perceive, interpret, and respond to the world around them in increasingly sophisticated ways.

Figure 2. The brain of future humanoids: "Neuromorphic computing" that challenges human brains

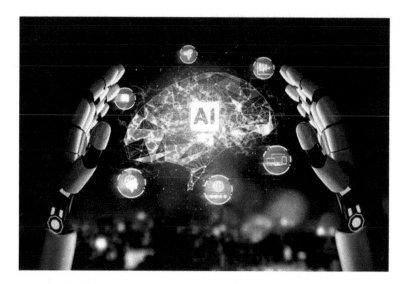

- **Perception:** AI enables humanoid robots to perceive and interpret sensory inputs from their environment, including visual, auditory, and tactile information.
- **Learning and Adaptation:** Through machine learning algorithms, humanoid robots can acquire new skills and knowledge from experience, allowing them to adapt to changing conditions and tasks.
- **Decision Making:** AI empowers humanoid robots to make autonomous decisions based on their perception and learned models, enabling them to navigate complex environments and execute tasks effectively.
- **Natural Language Processing:** AI-equipped humanoid robots are able to comprehend and produce natural language, enabling seamless communication with humans and facilitating interaction in various scenarios.
- **Emotion Recognition and Expression:** AI enables humanoid robots to recognize human emotions through facial expressions, tone of voice, and gestures, and respond appropriately with their own expressions and behaviours.

- **Social Interaction:** AI facilitates social interaction between humanoid robots and humans by enabling robots to interpret social cues, engage in conversation, and collaborate with humans in tasks.
- **Personalization:** AI allows humanoid robots to personalize their interactions with individuals based on their preferences, behaviours, and past interactions, enhancing the overall user experience.
- **Continuous Improvement:** Through iterative learning and feedback loops, AI enables humanoid robots to continuously improve their performance, refine their skills, and adapt to evolving user needs and expectations.

These roles collectively demonstrate the critical role of AI in enhancing the capabilities of humanoid robots and enabling them to effectively interact with humans and their environment.

Healthcare Applications of AI and Humanoid Robotics

Healthcare could be revolutionised by AI and humanoid robotics in a number of ways, including better patient outcomes, streamlined procedures, and higher overall standards of care.

Figure 3. AI and robotics in health care

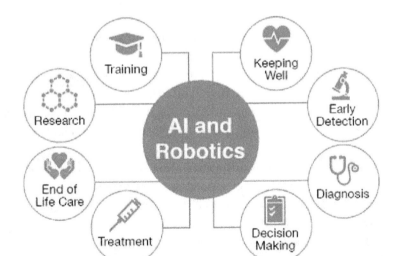

Humanoid robotics and artificial intelligence have various uses in the medical field.

Diagnosis and Medical Imaging: Radiologists can detect anomalies, tumours, or other medical issues more accurately and efficiently by using AI-powered algorithms to analyse medical images such as X-rays, MRIs, and CT scans. This may result in early illness diagnosis and treatment.

Personalized Medicine: Treatment plans and pharmaceutical regimens can be customised for each patient by using AI algorithms that can evaluate enormous volumes of patient data, including genetic information, medical histories, and lifestyle factors. This can reduce side effects and increase therapeutic efficacy.

Remote Patient Monitoring: Outside of conventional healthcare settings, patients' vital signs, activity levels, and other health parameters can be continuously monitored by wearable technology and sensors driven by artificial intelligence. This lowers the number of readmissions to hospitals and enhances patient care by enabling the early detection of health problems and prompt interventions.

Robot-Assisted Surgery: Humanoid robots equipped with advanced surgical tools can support surgeons during minimally invasive processes, enhancing accuracy and dexterity while minimising the menace of complications. Additionally, robots that can be operated by teleoperation can allow doctors to operate at a distance, increasing access to specialised care.

Rehabilitation and Physical Therapy: Humanoid robots with AI capabilities can assist patients with rehabilitation exercises and physical therapy routines, providing personalized guidance and feedback to optimize recovery from injuries or surgeries.

Medication Management: AI-powered systems can help healthcare providers optimize medication regimens by analyzing patients' medical histories, current medications, and potential drug interactions. This can lower the chance of negative drug effects and increase medication adherence.

Virtual Health Assistants: Chatbots and virtual assistants driven by artificial intelligence (AI) can offer patients individualised health advice, respond to inquiries about health, make appointments, and even enable remote consultations with medical professionals. This raises patient participation and facilitates better access to healthcare services.

Healthcare Operations and Administration: By forecasting patient admission rates, maximising personnel numbers, and automating administrative duties like scheduling and billing, artificial intelligence (AI) algorithms can optimise operations in healthcare facilities. Both efficiency and healthcare costs are decreased as a result.

Mental Health Support: Cognitive behavioural therapy (CBT), mood monitoring tools, and counselling are among the mental health services that AI-powered chatbots and virtual therapists can offer. These tools can supplement traditional therapy and provide continuous support to individuals with mental health conditions.

Elderly Care: Robots that are humanoid and have AI skills can help the elderly individuals with daily tasks, medication reminders, and companionship. These robots can help address the growing demand for elderly care services and alleviate caregiver burden.

Case Study

Ozturkcan, 2021 in his study presented the future of health care using Humanoid Service Robots during COVID 19 pandemic.

An overview of the use of robots in healthcare in pre- and post-pandemic scenarios is given by this case. Humanoid service robots are the subject of special attention because of their advantageous shape, size, and movement in human-designed physical locations. An assortment of instances from hospitals across the globe is showcased to illustrate the COVID-19 pandemic's utilisation of humanoid service robots in healthcare. The targeted future directions are intended to support improved policy and decision-making, which may reduce anxiety in people and increase acceptance. In addition to helping with healthcare during the pandemic, robots also helped with contactless check-in, support security at airports etc. Figure 4 highlights the statistics of the global use both airborne and ground robots for COVID-19 as of April 2020.

Figure 4. Global use of ground and aerial robots for COVID-19 as of April 2020

16 countries	7 countries	7 countries	5 countries	3 countries
Public Safety, Public Works, Non-Clinical Public Health	Clinical Care	Work, Critical Infrastructure, Quality of Life	Laboratory and Supply Chain Automation	Non-Hospital Care
Quarantine enforcement	Disinfecting point of care	Delivery	Delivery	Delivery to quarantined
Disinfecting public spaces	Healthcare worker telepresence	Socializing	Handling infectious materials	Quarantine or nursing home socializing
Identification of infected	Prescription/meal dispensing	Tele-commerce	Manufacture or Decon PPE	Off-Site Testing
Public service announcements	Patient intake & visitors	Robot assistants	Laboratory automation	Testing and care in nursing homes
Monitoring traffic flow	Patient and family socializing	Protection critical Infrastructure		

(Left axis: Decreasing frequency of reported uses)

Humanoid SR has becoming more widely used in healthcare because of the COVID-19 outbreak. The unexpected rise in healthcare demand fostered a culture where assistance of any kind was welcomed. Furthermore, the potential to reduce or completely avoid patient interaction created a chance to prevent the virus from infecting the valuable healthcare staff. Additionally, the perceived scarcity of personal protective equipment (PPE) was reduced. Physicians and nurses were able to spend more time with their patients by not having to worry about wearing and taking off personal protective equipment (PPE), which also required a time-consuming process. Humanoid SR assisted in re-establishing contact between patients and their families when hospital visits were no longer an option in order to provide comfort and support.

Using a tablet inserted in its chest, a humanoid SR in India by the name of Mitra helps COVID-19 patients in hospital beds communicate with their loved ones (SCMP, 2020). Mitra's human colleagues emphasise how troublesome the extended recuperation period and hospital visitation prohibitions are, particularly in the case of COVID-19. Patients also say that before Mitra assisted them, the loneliness in the hospital was unpleasant and that they felt much better. When they most need their support system, they can chat to their friends and family. In addition, Mitra can take a patient's temperature and, if necessary, help them schedule a consultation with a psychiatrist. Mitra has a facial recognition feature that allows it to identify individuals it has already encountered, allowing it to independently identify the patients. Additionally, Mitra helps with remote consultations as well, namely in cases when it shields the professionals who are at danger from any infections. In order to provide patients in various hospital areas with essential supplies such food and medication, another humanoid SR was also deployed (AlJazeera, 2020).

In the subsequent suit, Mitra was also hired in Bangalore (Fortis, 2020) to inspect every person entering a hospital. Mitra checks medical and non-medical personnel, including doctors, nurses, and other medical workers, for COVID-19 symptoms including fever and cough using facial and speech recognition technology. The assistance of Mitra in identifying and referring patients allows the human healthcare providers to keep a safe distance from the ill patients.

Consequences and Prospective Paths

Global healthcare is confronted with both immediate and long-term challenges, such as shifting populations, rising quality standards, resource shortages, and budgetary restrictions. Some of these problems can be solved with robots, especially with humanoid SR. In order to deliver better support and care in the not too distant future, the healthcare system will need to be redesigned, perhaps incorporating more humanoids in various roles. Enhancing comprehension of the causes, consequences, and implications of humanoid SR in healthcare might help make better decisions and policies that reduce people's anxiety and promote greater acceptance. Furthermore, future legislation that protects and fosters human-humanoid connection may influence how welcoming different participants are in common areas intended for collaboration. Stated differently, the objective ought to be the development of morally sound rules that investigate the potential benefits of humanoid SR for society. Informed ethical human-humanoid contact and proactive policy-making could then benefit the next generation of humanoid super intelligence.

Overall, AI and humanoid robotics hold immense promise for transforming healthcare by enhancing diagnosis, treatment, patient monitoring, and healthcare delivery while improving efficiency and reducing costs. However, ethical considerations and regulatory frameworks must be carefully considered in order to guarantee the fair and appropriate use of new technologies in healthcare environments.

Role of AI and Humanoid Robotics in Enhancing Human-Machine Interaction

- **Facial and Gesture Recognition:** Humanoid robots equipped with AI algorithms can recognize facial expressions and gestures, allowing them to perceive and respond to human emotions and intentions.
- **Natural Language Understanding:** AI-powered humanoid robots can interpret and respond to human speech, facilitating seamless communication and interaction.
- **Adaptive Behaviour:** AI enables humanoid robots to adapt their behavior and responses based on the context of interaction, enhancing their ability to engage with humans in various situations.
- **Personalization:** AI allows humanoid robots to personalize interactions by learning from past interactions and adapting to individual preferences and needs, fostering a more tailored and engaging user experience.
- **Social Skills Development:** Through AI, humanoid robots can assist in social skills development by providing feedback, guidance, and practice opportunities in social interactions.
- **Assistive Capabilities:** AI-powered humanoid robots can provide assistance to individuals with disabilities or special needs, facilitating daily tasks and improving accessibility to technology and services.
- **Entertainment and Education:** AI-driven humanoid robots can entertain and educate users through interactive storytelling, games, and educational activities, enhancing engagement and learning outcomes.
- **Collaborative Work:** Humanoid robots with AI capabilities can collaborate with humans in various tasks, such as manufacturing, research, and healthcare, augmenting human capabilities and productivity.
- **Accessibility and Inclusion:** AI and humanoid robotics contribute to creating more inclusive environments by providing assistance to individuals with disabilities or special needs. These technologies help bridge the gap between individuals with different abilities and enable them to participate more fully in society.

Overall, the amalgamation of AI and humanoid robotics in human-machine interaction is transforming how we interact with technology, making interactions more intuitive, personalized, and engaging. As these technologies continue to evolve, they hold immense potential to enhance various aspects of our daily lives and drive innovation across industries.

Role of AI and Humanoid Robotics in Advancing Automation

A new era of automation and human-machine interaction has been brought about by the union of AI and robots. AI-powered robots are revolutionising a number of industries, increasing productivity, and boosting human life quality thanks to recent advancements in computer vision, natural language processing, autonomous systems, and other areas.

Understanding robotics and artificial intelligence can be a challenging and complex process, but it can also be an exciting opportunity to learn about cutting edge technology (8). Artificial intelligence and robotics are two rapidly evolving sciences that provide a window into the future of automation and human-computer interaction. New advancements in both domains are made on a daily basis. The fields of artificial intelligence and robotics are expanding quickly, and new developments in these fields occur daily as these technologies offer an early look at automation and human-computer interaction in the future.

Because AI and robotics have the potential to revolutionise a variety of industries by increasing productivity, accuracy, and efficiency, we need them. These technologies lower hazards and human mistake by handling complicated data analysis, working in dangerous areas, and taking on repetitive jobs. Robotics and artificial intelligence (AI) have the potential to alleviate labour shortages in some industries and open up new opportunities in manufacturing, logistics, healthcare, and other fields. In the end, their incorporation into automation results in higher living standards, more innovation, and greater economic expansion.

The Latest AI Developments in Automation and Robotics

Advances in robotics and AI hold promise for modernising numerous industries through increased efficiency, reduced mistakes, and streamlined procedures. They also raise important moral and social concerns, like how automation will affect jobs and the need for strict regulations and safety measures as it becomes more commonplace. These are a few noteworthy developments in automation and robotics.

Neural Networks and Deep Learning

Automation has improved because to deep learning techniques like recurrent neural networks (RNNs) and convolutional neural networks (CNNs), which allow machines to analyse and comprehend enormous volumes of data. This has applications in picture and audio recognition, natural language processing, and other domains.

Computer Vision

Robots are now able to recognise and interact with their surroundings more successfully thanks to advancements in computer vision algorithms, which are becoming more precise and effective. This is important in fields like manufacturing, where robots are able to precisely recognise and operate objects.

Processing of Natural Language (NLP)

Human-robot interactions have improved as a result of NLP advancements. The efficacy of customer service and information retrieval has increased with the introduction of chatbots and virtual assistants that can understand and respond to natural language requests.

Learning via Reinforcement

Robots may now learn things by making mistakes and trying again thanks to reinforcement learning. Robots are becoming more adaptable in complicated contexts like logistics and autonomous vehicles because they can optimise their activities and independently adapt to new circumstances.

Automation of Robotic Processes (RPA)

RPA is being utilised more and more in corporate operations to automate jobs that follow rules and are repetitive. Data entry, data extraction, and other mundane jobs can be completed by bots, freeing up human workers to focus on more strategic and creative work.

Driverless Automobiles

One well-known use of robots and AI in automation is the creation of self-driving automobiles and trucks. These cars navigate and improve driving safety and efficiency through the use of cutting-edge sensors, machine learning algorithms, and real-time data.

Smart Manufacturing and Industry 4.0

Automation systems and robotics driven by AI are essential to the idea of Industry 4.0. IoT sensors, AI analytics, and robotics are used in "smart factories" to streamline production, cut down on downtime, and enhance quality assurance.

Automation in Agriculture

AI-powered drones and robots are being employed in agriculture to perform activities including crop monitoring, planting, and harvesting. Crop yields are raised and resource utilisation is optimised as a result.

Automation of Warehouse and Logistics

Robots and AI are being used by logistics organisations and e-commerce businesses to automate warehouse processes. This comprises product-picking, packing, and shipping robots and autonomous drones, which improve the efficiency of supply chains.

Robotics has mechanised many industries through the use of artificial intelligence and industrial-grade robots. Thanks to developments in robotics and automation, which have mechanised time-consuming, resource-intensive, and difficult operations and transferred them to machines that can work consistently and reliably, future sectors have experienced a transformation.

Intelligent robots are used in the aerospace, medical, and manufacturing sectors. They acquire machine learning training so they can work in quality control, automobile manufacturing, autonomous vehicle operation, and medical assistance. These robots are more accurate and productive than humans in doing tasks faster, even in dangerous environments.

Role of AI and Humanoid Robotics in Personalizing User Experiences

- **Tailored Recommendations:** AI-powered humanoid robots analyse user preferences and behaviour to offer personalized recommendations, whether in product choices, entertainment options, or educational materials.

Figure 5. Personalized recommendations

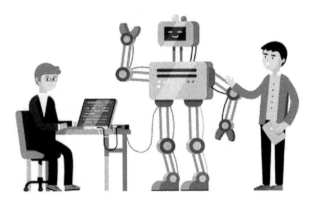

- **Adaptive Interactions:** Humanoid robots equipped with AI algorithms dynamically adjust their interactions based on individual user responses, ensuring a personalized and engaging experience.
- **Customized Services:** AI enables humanoid robots to provide customized services to users, such as personalized assistance in healthcare, education, or retail settings, catering to specific needs and preferences.
- **Emotional Recognition:** Humanoid robots with AI capabilities can recognize and respond to human emotions, allowing for empathetic interactions and personalized support based on the user's emotional state.
- **Learning and Adaptation:** AI-driven humanoid robots continuously learn from user interactions, adapting their behavior and responses over time to better meet individual needs and preferences.

Figure 6. AI-generated characters for supporting personalized learning and well-being

User interacting with an AI generated character during a health checkup

User learning from an AI generated character resembling an expert giving a lecture

Users becoming AI generated characters resembling historical characters for role playing

- **Assistive Capabilities:** AI and humanoid robotics personalize user experiences by providing tailored assistance to individuals with disabilities or special needs, ensuring accessibility and inclusivity in various contexts.
- **Context Awareness:** AI-powered humanoid robots utilize context-awareness to understand the user's environment and situation, delivering relevant and timely information or assistance tailored to the specific context.
- **Feedback and Improvement:** Through AI analytics, humanoid robots gather user feedback to continuously improve their performance and enhance the personalized user experience over time.

Role of AI and Humanoid Robotics in Higher Education

AI and humanoid robotics are increasingly playing significant roles in higher education, revolutionizing the learning experience for both learners and teachers. Here are some of the key roles they play:

- **Personalized Learning**: To tailor the learning experience for each student, AI systems can examine their performance statistics, learning preferences, and learning habits. This improves learning outcomes by allowing teachers to modify their lesson plans and instructional strategies to meet the needs of each unique student.
- **Virtual Teaching Assistants**: AI-powered virtual teaching assistants can help students around-the-clock by responding to inquiries, offering feedback on tasks, and providing individualized instruction (Rajagopal, 2018). This reduces the burden on educators and enables students to receive immediate assistance whenever they need it.

Figure 7. A student uses the chat box to ask Maria a question

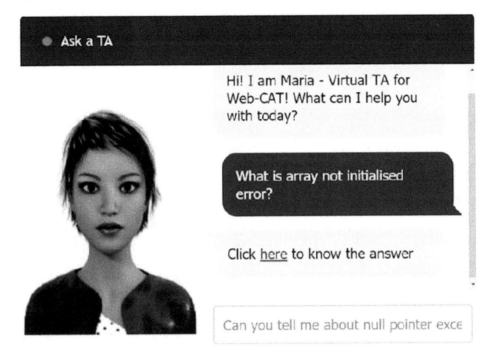

- **Improved Teaching Tools**: AI-powered teaching resources can help teachers develop dynamic and interesting lesson plans. For example, AI algorithms can generate personalized quizzes, simulations, and educational games that cater to different learning styles and preferences.
- **Research and Data Analysis**: AI can aid researchers in higher education through examining enormous volumes of data, seeing trends, and producing insights. This can accelerate the pace of research and lead to new discoveries across various disciplines.
- **Humanoid Robotics for Hands-On Learning**: Humanoid robots can be used in higher education to provide hands-on learning experiences in fields such as robotics, engineering, and computer science. Students can interact with these robots, program them, and gain practical skills that are essential for their future careers.
- **Language Learning and Communication**: AI-powered language learning platforms can help students improve their language skills through personalized exercises, interactive conversations, and real-time feedback. Additionally, humanoid robots equipped with natural language processing capabilities can serve as language tutors, providing immersive language learning experiences.

- **Accessibility and Inclusivity**: AI technology can contribute to increasing the inclusivity and accessibility of higher education for individuals with disabilities. For example, AI-powered captioning and translation tools can improve the learning experience for students with hearing impairments or those who speak different languages.
- **Predictive Analytics for Student Success**: In order to identify at-risk kids who might require extra support or intervention, AI systems can analyse student data. Teachers can increase student success rates and retention by implementing tailored interventions based on predictive data about student outcomes and behaviour.

Case Study: An online teaching assistant to help students in their programming endeavors.

Learning programming can be challenging for inexperienced programmers. They struggle to grasp the subject and have a variety of misconceptions. For many students, programming can be an unpleasant experience all around. They can believe that programming is above their capabilities and feel alone in the programming community. Automated grading technologies are widely used in schools to evaluate student work and give them early feedback.

The author of this case study, Rajagopal (2018), introduced Maria, a virtual teaching assistant who will live inside Web-CAT and assist students in easing some of their negative feelings towards programming. We have used an animated human-like character, known as pedagogical agent, for Maria as it is widely use in pedagogy to help students. This project's main objective is to give students enough emotional support so they can overcome some of the anxiety associated with Java programming assignments and projects. The goal is to provide the pupils with assistance in manners that a human teaching assistant could. Although we think the present Web-CAT system gives students insightful feedback, there aren't enough resources to help them emotionally, which could have an impact on their behaviour.

This research concentrates on the choice of selection of the pedagogical agent, the chatbot's implementation, and the technology that powers it. By definition, pedagogical agents simulate human-like interfaces. Thus, an animated avatar serves as the implementation of a pedagogical agent as shown in Fig.6. The implementation of chatbots is the study's most significant contribution. It was decided to also install Maria as a chatbot. The chat window that was created is seen in Fig.6, conversation window featuring Maria's animated avatar is located on the left, while the chat history is located on the right. Maria opens the chat window by introducing herself. JavaScript, HTML, and CSS are examples of front-end technologies used in the chat box's implementation. Students can use the given text box to type their questions and hit the Enter key. As demonstrated in Fig 6 the user's query and Maria's response are then shown in the chat history.

Architecture/Methodology

The chatbot is a part of WebCAT, which is the front-end application. In order to make it easier for front-end and back-end technologies to communicate, a middle layer serves as middleware. Maria is powered by Chat Script, which runs on a CentOS server as the back-end. The middle layer consists of an Apache server with a PHP script running on it. The PHP script will listen for AJAX requests from Web-CAT and send the requests over raw sockets to the back end. Now, Chat Script can only listen to raw TCP socket conversation, therefore this intermediary layer was employed to get past that problem. Web-CAT does not enable raw socket programming because we utilise plain JavaScript for client side programming. However, PHP is a strong scripting language that facilitates programming with raw sockets. Therefore, our approach will send AJAX requests from Web-CAT to the Apache server, which the PHP script will then accept and transfer to Chat Script using raw socket programming. Conversely, Chat Script will use raw sockets to transmit its response to the PHP script, which will then use an AJAX response to deliver the response back to Web-CAT as shown in Figure 8.

Figure 8. Chatbot's architecture

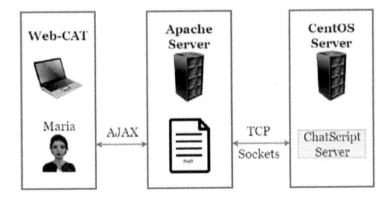

Students Using the Chatbox

Let's look at an example where students send in a code to Web-CAT and "chat" with Maria regarding mistakes. Typically, students use the IDE of their choosing to code when they start an assignment involving programming. They will write their code in accordance with the instructor's specifications. They must submit the code to Web-CAT before they can view the feedback from Web-CAT. Figure 9 displays a few compiler problems made by the learner and an expanded coding area with thorough feedback. Figure 10 illustrates a mistake in the extended part, together with the justification next to it and Figure 11 shows a popup window with Maria's comment to the "expected" error.

Figure 9. Few compiler problems made by the learner and an expanded coding area with thorough feedback

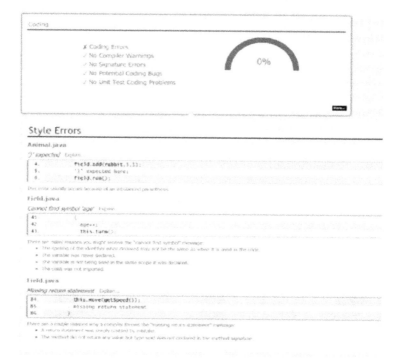

Figure 10. The feedback page in the enlarged section has the explanation link next to the error messages

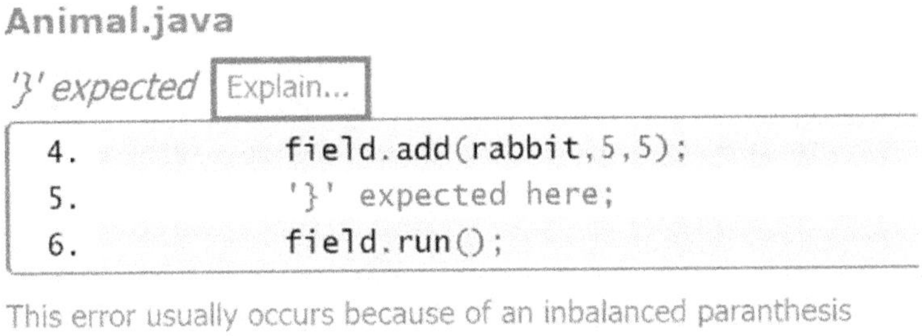

Animal.java

'}' expected | Explain... |

```
4.              field.add(rabbit,5,5);
5.                 '}' expected here;
6.              field.run();
```

This error usually occurs because of an inbalanced paranthesis

Figure 11. A popup box containing the response from Maria about the '} expected' error

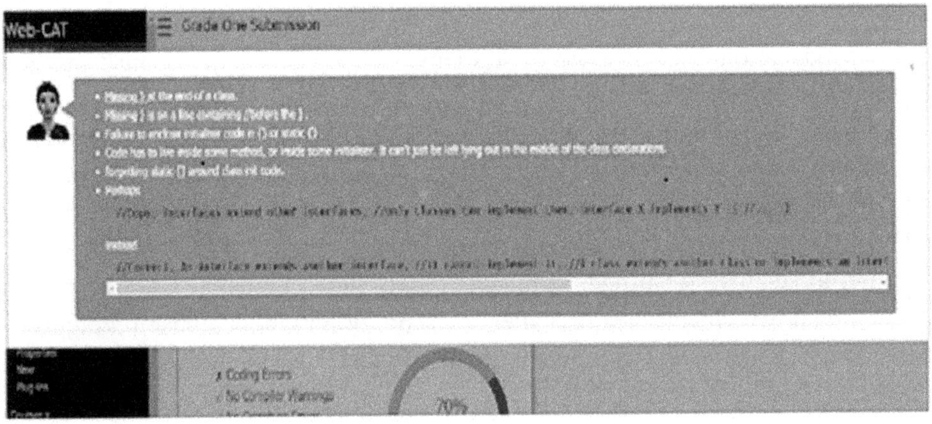

CONCLUSION

The creation and application of Maria a virtual teaching assistant in Web-CAT, to provide students with emotional support during their programming tasks has been defined and discussed in this research. It outlined the current issues that students go into when completing projects and gave the reasons why this initiative was necessary. This study showed on how students find programming difficult, how pedagogical agents work well in classroom settings, and how best to employ them to inspire students. The study gave a demonstration of the User Interface so that students could see how Maria will interact with them in various settings. The study evaluated Maria's effectiveness and discussed our evaluation strategy.

Overall, AI and humanoid robotics have the ability to completely change higher education by increasing accessibility, personalisation, and engagement while also giving teachers the tools they need to improve student outcomes and efficacy. But it's crucial to address privacy issues, ethical issues, and make sure these technologies are used in educational settings in a responsible and ethical manner.

Role of AI and Humanoid Robotics Pushing Boundaries in Addressing Societal Challenges

Healthcare Support: AI-powered humanoid robots assist in healthcare by providing personalized care to aging populations, monitoring vital signs, and assisting in rehabilitation, addressing the challenges of an aging society and improving quality of life.

Education Accessibility: Humanoid robots equipped with AI algorithms support education by providing personalized tutoring, adaptive learning experiences, and access to educational resources in remote or underserved areas, addressing disparities in education accessibility.

Disaster Response: AI-driven humanoid robots aid in disaster response by conducting search and rescue missions, navigating through hazardous environments, and providing assistance to rescue teams, addressing challenges in disaster management and emergency preparedness.

Mental Health Assistance: Humanoid robots with AI capabilities offer support to individuals with mental health challenges by providing companionship, cognitive-behavioral therapy exercises, and emotional support, addressing the growing need for mental health services.

Environmental Conservation: AI-powered humanoid robots contribute to environmental conservation efforts by monitoring ecosystems, collecting data on wildlife populations, and assisting in habitat restoration projects, addressing challenges in biodiversity conservation and climate change mitigation.

Accessibility and Inclusion: Humanoid robots with AI algorithms promote accessibility and inclusion by assisting individuals with disabilities or special needs in daily tasks, facilitating social interaction, and improving access to technology and services, addressing challenges in accessibility and inclusivity.

Social Isolation: AI-driven humanoid robots combat social isolation by providing companionship, social interaction, and emotional support to individuals who may be isolated or lonely, addressing challenges in mental well-being and social connectedness.

Economic Empowerment: Humanoid robots equipped with AI capabilities create opportunities for economic empowerment by enhancing productivity, creating new jobs in robotics-related industries, and supporting workforce development, addressing challenges in unemployment and economic inequality.

Figure 12. Role of AI and humanoid robotics pushing boundaries in addressing societal challenges

Role of AI and Humanoid Robotics in Natural Language Processing

The amalgamation of AI and humanoid robotics in natural language processing (NLP) has not only revolutionized human-robot interaction but also created fresh opportunities for innovation and application. In addition to understanding and responding to human language, AI-powered humanoid robots equipped with NLP capabilities can now analyze sentiment, detect intent, and generate more contextually relevant responses. This deeper understanding of language nuances allows for more meaningful and engaging interactions between humans and robots. Furthermore, the advancements in NLP have facilitated the development of conversational agents and chatbots that can support users in a variety of duties, from scheduling appointments and providing customer support to teaching languages and offering personalized recommendations. These AI-driven applications have the potential to streamline processes, improve efficiency, and enhance user satisfaction across various industries.

Figure 13. AI in natural language processing

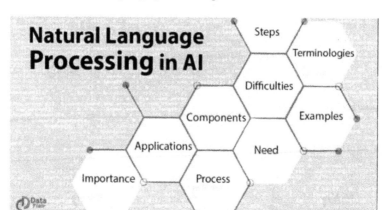

Moreover, the combination of AI and humanoid robotics in NLP has implications for language learning and education. By providing interactive language learning experiences, humanoid robots can help individuals improve their language skills in a more immersive and engaging manner. Additionally, in educational settings, AI-powered robots can serve as interactive tutors, offering personalized learning experiences tailored to each student's needs and abilities.

Overall, the integration of AI and humanoid robotics in NLP not only enhances human-robot interaction but also brings about transformative changes in various sectors, including customer service, education, and language learning. We may anticipate much more advanced NLP uses in humanoid robotics as technology develops, which will further obfuscate the distinction between humans and machines and change how we interact and communicate.

Role of AI and Humanoid Robotics in Multi-Modal Interaction

The synergy between artificial intelligence (AI) and humanoid robotics has propelled the development of multi-modal interaction, revolutionizing how robots perceive and engage with humans across various sensory modalities. Through sophisticated AI algorithms, humanoid robots can seamlessly integrate inputs from diverse sources such as vision, speech, touch, and proprioception, enabling a holistic understanding of their environment and interactions. AI-powered speech recognition and synthesis capabilities empower robots to comprehend spoken commands and respond with natural language, fostering fluid communication. Additionally, AI-driven gesture and facial expression recognition allow robots to interpret non-verbal cues, enhancing their ability to empathize and engage with users. By harnessing haptic feedback mechanisms, humanoid robots can physically interact with users, further enriching the interaction experience. AI facilitates adaptive interaction, enabling robots to adjust their behaviour based on contextual cues and user preferences, leading to personalized and engaging interactions. Overall, the amalgamation of AI and humanoid robotics in multi-modal interaction holds immense potential to redefine human- robot collaboration and communication, paving the way for more intuitive and immersive interactions in various domains.

Furthermore, the amalgamation of AI and humanoid robotics in multi-modal interaction extends beyond mere sensory perception. AI algorithms enable robots to analyze and understand the context of human interactions, allowing them to adapt their responses dynamically. This contextual awareness enables humanoid robots to anticipate user needs, provide proactive assistance, and offer relevant information in real-time, enhancing the overall user experience.

Moreover, the synergy between AI and humanoid robotics in multi-modal interaction facilitates the development of assistive technologies for individuals with disabilities. By combining sensory inputs with intelligent processing capabilities, humanoid robots can assist users with tasks such as navigation, communication, and daily living activities, empowering them to lead more independent and fulfilling lives.

Additionally, the integration of AI and humanoid robotics opens up new avenues for collaborative and cooperative tasks in various industries. Humanoid robots equipped with multi-modal interaction capabilities can work alongside humans in environments such as manufacturing, healthcare, and retail, complementing human skills and improving overall productivity and efficiency.

Overall, the seamless integration of AI and humanoid robotics in multi-modal interaction represents a significant step forward in human-robot collaboration and communication. As technology continues to advance, we can expect even greater strides in creating more intuitive, empathetic, and immersive interactions between humans and robots, leading to transformative changes across diverse domains and industries.

Role of AI and Humanoid Robotics Pushing Boundaries in Research and Innovation

- **Cognitive Modelling:** Humanoid robots with AI capabilities serve as platforms for studying human cognition, enabling researchers to develop and test cognitive models, theories, and hypotheses in real-world environments.
- **Cross-disciplinary Collaboration:** AI and humanoid robotics foster collaboration between diverse fields such as computer science, neuroscience, psychology, and engineering, leading to innovative solutions and interdisciplinary breakthroughs.

- **Human-Robot Interaction:** AI-driven humanoid robots push boundaries in understanding human-robot interaction, leading to advancements in interface design, communication protocols, and social robotics applications.
- **Adaptive Learning:** AI algorithms enable humanoid robots to learn from experience and adapt to new tasks and environments, driving research in lifelong learning, transfer learning, and continual adaptation.
- **Autonomous Exploration:** Humanoid robots equipped with AI explore unstructured environments, such as space, deep sea, or disaster zones, pushing boundaries in autonomous exploration, mapping, and navigation.
- **Ethical and Social Implications:** Research at the intersection of AI and humanoid robotics explores ethical and social implications, including questions of robot rights, human-robot relationships, and the impact of automation on society.
- **Bio-inspired Design:** AI-driven humanoid robotics draws inspiration from biology and nature, leading to innovative designs and functionalities inspired by biological systems, such as locomotion, sensing, and manipulation.
- **Future Technology Applications:** Research in AI and humanoid robotics anticipates future technology applications, such as smart cities, autonomous vehicles, and robotic companions, driving innovation and shaping the future of technology and society.

Figure 14. Data processing in research

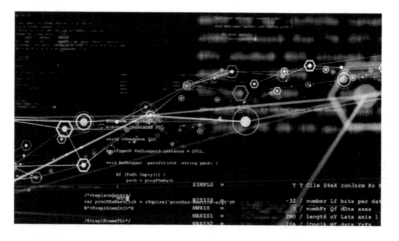

AI-Powered Military Robots: Boosting Defence Power

AI-driven robots have being extensively investigated by the military to improve defence capabilities. AI military robots could lower human risk by carrying out crucial jobs in difficult and hazardous locations. These robots can be employed for risky tasks including bomb disposal, surveillance, and reconnaissance (Agarwal, 2023). Military robots are now able to make intelligent choices based on data collected in real time thanks to developments in AI algorithms, machine learning, and computer vision. This increases their operational efficiency and efficacy in combat.

Figure 15. Types of military robots

The integration of AI in military applications has been a significant trend in modern warfare, revolutionizing various aspects of defence and security. Here are some key areas where AI is making an impact:

Autonomous Systems: Drones, unmanned ground vehicles (UGVs), and unmanned aerial vehicles (UAVs) are examples of AI-powered autonomous systems that are being used more and more for combat operations, reconnaissance, and surveillance. These systems can operate independently or in conjunction with human operators, reducing the risk to human life in dangerous situations.

Predictive Maintenance: AI algorithms are used for military machinery and equipment predictive maintenance. Artificial Intelligence (AI) can forecast when components are likely to break by analysing data from sensors and other sources. This enables proactive maintenance to be carried out, reducing downtime and guaranteeing operational readiness.

Cyber Defense: AI is essential for improving cybersecurity because it can quickly detect and neutralise cyber threats. Large volumes of data can be analysed by AI algorithms, which can then be used to spot trends, spot possible cyber attacks, and take quick action to lessen the harm.

Logistics and Supply Chain Management: AI optimization algorithms are used to improve logistics and supply chain management in the military. These algorithms help in optimizing routes, managing inventory, and predicting demand, thereby ensuring efficient distribution of resources and minimizing costs.

Strategic Decision Making: Artificial intelligence (AI)-driven decision support systems analyse enormous volumes of data from numerous sources, including satellite images, intelligence reports, and historical data, to help military leaders make well-informed judgements. These systems provide insights and recommendations to aid in strategic planning and execution.

Training and Simulation: AI is utilized in military training and simulation environments to create realistic scenarios and adaptive training programs. When mixed with artificial intelligence (AI), virtual reality (VR) and augmented reality (AR) technologies provide immersive training experiences that improve military personnel's abilities and readiness.

Target Recognition and Tracking: AI algorithms enable automatic target recognition (ATR) and tracking systems, which can identify and track objects of interest such as enemy vehicles, aircraft, or personnel. These systems enhance situational awareness and enable more accurate and timely engagement of targets.

Medical Support: Military healthcare providers benefit from AI-powered medical technologies that help with injury diagnosis, patient outcome prediction, and treatment plan optimisation. These systems can deliver individualised medical care to military personnel in distant or harsh situations by analysing medical data, such as imaging scans and patient records.

Advanced Military Robots with Various Applications (Anuradha & Janith, 2023)

SAFFiR

The military robot SAFFiR (Shipboard Autonomous Firefighting Robot) measures five feet ten inches tall and weighs sixty-five kilogrammes. The robot, created by Virginia Tech academics, is technologically advanced and intended to put out flames that occur on military vessels. Although SAFFiR requires a chain to function, it is capable of measuring steps and operating a fire hose. It can travel in a variety of areas thanks to its superhuman range of motion, which is made possible by its special mechanism plan. The ultimate objective of SAFFiR is to improve futuristics capabilities while collaborating with Navy officers rather than replacing them.

Figure 16. Ship board autonomous fire fighting robot

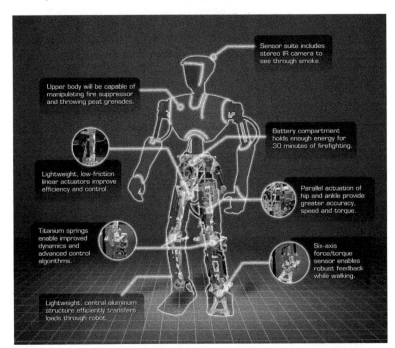

MUTT (Multi-Utility Tactical Transport)

There are two types of MUTT, an unmanned ground vehicle: tracked and wheeled.

By reducing the amount of gear the warriors must carry when traversing challenging terrain on foot, MUTT travels with them, simplifying their journey. There are three sizes available for this autonomous battle vehicle: 6x6, 8x8, and tracked. Standing 112 inches long and 60 inches wide, the 8×8 MUTT can accommodate up to 544 kg. With just one petrol tank, it can travel up to 97 km and produce up to 3,000 watts of power. The MUTT project's research is looking into this robot's potential future features.

Figure 17. Multi-utility tactical transport robot vehicle

MORAL CONUNDRUMS IN AI-POWERED MILITARY ROBOTS

Autonomy and Human Control in Balance

A primary moral conundrum with AI military robots is striking the correct balance between autonomy and human oversight. Although the freedom to make decisions on one's own can increase productivity and performance, total autonomy creates questions about responsibility and possible unexpected effects. Maintaining a balance between machine autonomy and human monitoring is essential to ensuring the ethical and acceptable application of AI in military operations.

The use of artificial intelligence (AI) in military robots prompts worries about the possible effects of cutting-edge technology in combat.

There are concerns because AI makes decisions more quickly, is more harmful, and involves fewer people in the process, conflict may escalate. Strong laws, rules, and international agreements are needed to address these issues in order to guarantee responsible usage of AI and reduce the possibility of unforeseen repercussions.

The development, application, and deployment of AI in military robots must be governed by precise policies and procedures that are established by military organisations. Adherence to ethical values, accountability, and transparency ought to be given top priority in these rules.

Additionally, to solve the significant ethical conundrums related to AI in military applications, continued research and discussion among engineers, ethicists, and policymakers are required.

Challenges in implementing AI and Humanoid Robotics

Technical Complexity: Developing AI algorithms and integrating them into humanoid robot platforms requires expertise in multiple disciplines, including robotics, computer science, and artificial intelligence. Ensuring the seamless interaction between hardware and software components poses technical challenges that need to be addressed.

Safety Concerns: As humanoid robots become more autonomous and interact closely with humans, ensuring their safety and reliability becomes paramount. Safety standards and protocols need to be established to mitigate risks associated with physical interaction, navigation in dynamic environments, and handling unexpected situations.

Ethical Considerations: AI-driven humanoid robots raise ethical questions related to privacy, autonomy, and accountability. Robots must be used responsibly and ethically, hence concerns like data privacy, algorithmic prejudice, and the effect of automation on employment must be addressed.

Human-Robot Interaction: It is difficult to design user-friendly and natural interfaces for human-robot interaction, especially when it comes to sensing and reacting to human intentions, emotions, and social cues. Advances in natural language processing, gesture recognition, and emotion detection are necessary to ensure human-robot collaboration and communication.

Adaptability and Learning: Humanoid robots need to adapt to diverse environments and tasks, requiring robust learning and adaptation capabilities. Developing algorithms that enable robots to learn from experience, generalize knowledge, and adapt to new situations is a significant challenge in AI and robotics research.

Cost and Accessibility: The development and deployment of AI-powered humanoid robots involve significant costs, limiting accessibility to advanced technologies for certain sectors or populations. Finding cost-effective solutions and guaranteeing fair access to humanoid robotics technologies are essential for widespread adoption and societal benefit.

Regulatory Frameworks: As AI and humanoid robotics technologies advance, regulatory frameworks need to keep pace to ensure compliance with safety, privacy, and ethical standards. Establishing regulations and guidelines for the design, deployment, and use of humanoid robots is crucial for ensuring societal acceptance and trust.

Researchers, legislators, industry stakeholders, and the general public must work together to find creative solutions, set moral standards, and promote the responsible application of AI and humanoid robots technologies in order to address these issues.

CONCLUSION

In conclusion, the amalgamation of applied AI with humanoid robotics holds immense promise for revolutionizing diverse aspects of human life, from healthcare and education to manufacturing and beyond. By harnessing the synergies between AI techniques and robotic platforms, we can unlock new possibilities for enhancing human-robot interaction, improving task performance, and addressing societal challenges. However, realizing this potential requires concerted efforts to overcome technical hurdles, ethical dilemmas, and regulatory frameworks, ensuring that AI-driven humanoid robots serve as enablers of positive change and contribute to the betterment of humanity.

Through applications across a range of domains such as healthcare, education, entertainment, and industry, AI-powered humanoid robots are increasing output, effectiveness, and user happiness. They assist with tasks ranging from customer service and care giving to language tutoring and industrial automation, driving technological innovation and addressing societal challenges.

However, challenges such as technical complexity, safety concerns, ethical considerations, and human-robot interaction pose significant hurdles to the widespread adoption and responsible deployment of AI and humanoid robotics technologies. Addressing these challenges requires interdisciplinary collaboration, regulatory frameworks, and ethical guidelines to ensure the beneficial integration of AI-driven humanoid robots into society.

REFERENCES

Agarwala, N. (2023). Robots and artificial intelligence in the military. *Obrana a Strategie (Defence and Strategy), 23*(2), 83–100. doi:10.3849/1802-7199.23.2023.02.083-100

Anuradha, J. (2023). *Used of Artificial Intelligence & Robotics in Military Field.* Academic Press.

Deo, N., & Anjankar, A. (2023). Artificial Intelligence with robotics in Healthcare: A narrative review of its viability in India. *Cureus.* Advance online publication. doi:10.7759/cureus.39416 PMID:37362504

Okagbue, E. F., Muhideen, S., Anulika, A. G., Nchekwubemchukwu, I. S., Chinemerem, O. G., Tsakuwa, M. B., Achaa, L. O., Adarkwah, M. A., Funmi, K. B., Nneoma, N. C., & Mwase, C. (2023). An in-depth analysis of humanoid robotics in Higher Education System. *Education and Information Technologies, 29*(1), 185–217. doi:10.1007/s10639-023-12263-w

Ozturkcan, S., & Merdin-Uygur, E. (2021). Humanoid Service Robots: The future of healthcare? *Journal of Information Technology Teaching Cases, 12*(2), 163–169. doi:10.1177/20438869211003905

Pérez, L., Rodríguez-Jiménez, S., Rodríguez, N., Usamentiaga, R., García, D. F., & Wang, L. (2019). Symbiotic human–robot collaborative approach for increased productivity and enhanced safety in the aerospace manufacturing industry. *International Journal of Advanced Manufacturing Technology, 106*(3–4), 851–863. doi:10.1007/s00170-019-04638-6

PricewaterhouseCoopers. (n.d.). *No longer science fiction, AI and Robotics Are Transforming Healthcare.* PwC. https://www.pwc.com/gx/en/industries/healthcare/publications/ai-robotics-new-health/transforming-healthcare.html

Rajagopal, M., & Babu, M.N. (2018). *Virtual Teaching Assistant to Support Students' Efforts in Programming.* Academic Press.

Reddy, S. (2023, October 26). *NLP in Robotics: Enhancing Human-Robot Interaction.* Pss Blog. https://www.pranathiss.com/blog/nlp-human-robot-interaction/

Tabassum, Z. (2023). Artificial intelligence and blockchain technology for secure smart grid and power distribution automation. In AI and Blockchain Applications in Industrial Robotics (pp. 226–252). IGI Global. doi:10.4018/979-8-3693-0659-8.ch009

Tong, Y., Liu, H., & Zhang, Z. (2024). Advancements in humanoid robots: A Comprehensive Review and future prospects. *IEEE/CAA Journal of Automatica Sinica, 11*(2), 301–328. doi:10.1109/JAS.2023.124140

Venkataswamy, R., Janamala, V., & Cherukuri, R. C. (2023). Realization of humanoid doctor and real-time diagnostics of disease using internet of things, Edge Impulse Platform, and chatgpt. *Annals of Biomedical Engineering, 52*(4), 738–740. doi:10.1007/s10439-023-03316-9 PMID:37453975

. Vrontis, D., Christofi, M., Pereira, V., Tarba, S., Makrides, A., & Trichina, E. (2023). Artificial Intelligence, robotics, Advanced Technologies and Human Resource Management: A systematic review. *Artificial Intelligence and International HRM,* 172–201. doi:10.4324/9781003377085-7

KEY TERMS AND DEFINITIONS

Artificial Intelligence: Artificial Intelligence (AI) refers to the development of computer systems or algorithms that can perform tasks that typically require human intelligence. These tasks include learning, reasoning, problem-solving, perception, understanding natural language, and interacting with the environment. AI systems aim to simulate cognitive functions associated with human minds, such as learning from experience, adapting to new situations, and making decisions based on available information. AI techniques and approaches include Machine Learning, Deep Learning, Nature Language Processing, Computer Vision and Robotics. AI has diverse applications across various industries and domains, including healthcare, finance, transportation, education, entertainment, and more. It has the potential to revolutionize how tasks are performed, enhance productivity, improve decision-making, and address complex challenges facing society.

Chat Bots: Chatbots are computer programs or AI systems designed to simulate human conversation through text or speech interactions. They use natural language processing (NLP) and machine learning algorithms to understand and respond to user queries, requests, or commands in a conversational manner. Chatbots can be deployed across various platforms, including websites, messaging apps, social media, and virtual assistants. The primary purpose of chatbots is to automate customer service, provide information, answer questions, assist with tasks, and engage users in personalized conversations. They can range from simple rule-based systems that follow predefined scripts to more advanced AI-powered models capable of learning from interactions and adapting their responses over time. Chatbots have become increasingly prevalent in business and customer service settings, helping companies improve efficiency, enhance user experience, and provide round-the-clock support. They can also be used for various other purposes, such as lead generation, sales, education, entertainment, and personal assistance.

Education: Education is the process of facilitating learning, acquisition of knowledge, skills, values, beliefs, and habits. It occurs through various formal and informal methods, such as classroom instruction, self-directed study, practical experience, and interaction with peers and mentors. Education aims to foster intellectual, social, emotional, and physical development in individuals, enabling them to adapt, grow, and contribute to society. It encompasses a wide range of subjects and disciplines, including mathematics, science, literature, history, arts, and physical education. Education systems vary across countries and cultures but generally include primary, secondary, and higher education levels, as well as vocational and lifelong learning opportunities. The ultimate goal of education is to empower individuals to lead fulfilling lives, participate actively in civic life, and pursue their aspirations and potential.

Entertainment: Entertainment refers to any activity, performance, or form of media that is designed to amuse, entertain, or engage an audience. It encompasses a wide range of experiences and content, including movies, television shows, music, theater, sports, video games, amusement parks, and live performances. The primary goal of entertainment is to provide enjoyment, relaxation, and diversion from daily routines or responsibilities. It can evoke emotions, spark imagination, and foster social interaction, serving as a means of cultural expression, storytelling, and artistic creativity. Entertainment plays a significant role in society by shaping values, influencing trends, and providing shared experiences that bring people together and contribute to their overall well-being and quality of life.

Generative AI: Generative AI refers to a subset of artificial intelligence techniques and algorithms designed to create new content, such as images, text, audio, and video, that resembles human-generated data. Unlike traditional AI systems that primarily analyze and process existing data, generative AI models are capable of generating novel content based on patterns and structures learned from large datasets during training. These models leverage techniques like neural networks, deep learning, and probabilistic modeling to generate realistic and diverse outputs.

Healthcare: Healthcare refers to the maintenance or improvement of health through the prevention, diagnosis, treatment, and management of illness, injury, and other physical and mental impairments in individuals or populations. It encompasses a broad range of services, including medical, dental, nursing, pharmacy, allied health, and public health, delivered by various professionals and institutions. Healthcare aims to promote overall well-being and quality of life by addressing health-related issues and providing medical interventions tailored to the needs of patients and communities.

Manufacturing: Manufacturing is the process of converting raw materials, components, or parts into finished goods through various techniques, tools, and machinery. It involves a series of steps, including design, production planning, fabrication, assembly, quality control, and distribution. Manufacturing can take place in various industries, such as automotive, electronics, aerospace, pharmaceuticals, and consumer goods. The goal of manufacturing is to create products that meet specific quality standards, cost targets, and customer requirements efficiently and effectively. It plays a crucial role in economic development by generating employment, driving innovation, and supporting the production of goods for consumption and export.

Robotics: Robotics is a multidisciplinary field that involves the design, construction, operation, and use of robots to perform tasks in various environments. Robots are autonomous or semi-autonomous machines that can be programmed to execute specific actions, often mimicking human or animal movements. The field of robotics encompasses aspects of mechanical engineering, electrical engineering, computer science, and artificial intelligence. It spans a wide range of applications, from industrial manufacturing and automation to space exploration, healthcare, agriculture, and entertainment.

Chapter 2
Harnessing Applied AI:
Transforming Industry and Business

Beulah Viji Christiana
Department of Master of Business Administration, Panimalar Engineering College, Chennai, India

B. Charith
Department of Management and Commerce, International Institute of Business Studies, Bengaluru, India

Krishnamohan Reddy Kunduru
https://orcid.org/0009-0009-8060-6216
Department of Engineering Design, Overhead Door Corporation, Lewisville, USA

M. Thangatamilan
Department of Electronics and Instrumentation Engineering, Kongu Engineering College, Perundurai, India

J. Chandrasekar
Department of Social Work, Madras School of Social Work, Chennai, India

A. Rajendra Prasad
Department of Mechanical Engineering, Sri Sai Ram Engineering College, Chennai, India

ABSTRACT

This chapter delves into the transformative potential of applied artificial intelligence (AI) in various industries, highlighting its role in optimizing processes across sectors like manufacturing, logistics, and healthcare. It also discusses how AI is reshaping business operations, from sales and marketing to customer service and human resources, through the adoption of AI-powered tools and platforms. The chapter provides a comprehensive overview of AI's impact on modern business ecosystems. The chapter discusses the use of AI to enhance customer experiences, focusing on personalized interactions, tailored recommendations, and improved satisfaction across industries. It also addresses ethical considerations, governance frameworks, and future trends in AI. The chapter highlights the transformative potential of AI and advocates for continuous adaptation and innovation to fully utilize its benefits in industry and business.

DOI: 10.4018/979-8-3693-2399-1.ch002

INTRODUCTION

Artificial Intelligence (AI) is revolutionizing various aspects of life, including industry and business operations. Applied AI, specifically, uses AI technologies to tackle real-world challenges and improve organizational performance. This introduction provides an overview of applied AI, its significance, and its diverse applications across industries. Unlike theoretical AI research, applied AI is driven by practical objectives and tangible outcomes, aiming to improve efficiency, productivity, and decision-making across diverse domains. The primary goal is to deliver tangible value (Suryadevara, 2023).

Applied AI is crucial for unlocking new opportunities and driving innovation in various industries. It uses machine learning, natural language processing, and computer vision to provide deeper insights, automate repetitive tasks, and make data-driven decisions at scale. This transformative potential has driven its adoption across sectors like manufacturing, finance, healthcare, and retail. AI is transforming industries like manufacturing, logistics, and supply chain management by enhancing process optimization. Predictive maintenance algorithms can identify equipment failures early, reducing downtime and productivity (Jarrahi et al., 2023). AI-powered inventory management systems optimize stock levels, enhancing resource utilization and customer satisfaction. These AI-driven solutions are increasingly being adopted in industries to streamline operations and reduce costs.

AI is revolutionizing traditional business functions like sales, marketing, and customer service by personalizing interactions, delivering targeted campaigns, and providing proactive support. Natural language processing algorithms enable sentiment analysis and translation, while AI-powered chatbots and virtual assistants offer round-the-clock assistance, enhancing customer satisfaction and loyalty. The adoption of applied AI faces challenges such as ethical considerations, data privacy concerns, algorithmic biases, talent acquisition, data quality, and infrastructure readiness (Zaki, 2019). To overcome these, a holistic approach involving technical expertise, ethical frameworks, and regulatory compliance is needed. The future of applied AI holds immense promise, with applications like autonomous vehicles, smart cities, personalized medicine, and predictive analytics. However, realizing this potential requires ongoing collaboration between industry stakeholders, policymakers, and researchers to address emerging challenges and seize innovation opportunities (Kitsios & Kamariotou, 2021).

Applied AI is a significant shift in how industries use technology to create value and gain a competitive edge. It allows organizations to optimize processes and provide better customer experiences. However, it's crucial to navigate ethical, regulatory, and technical complexities to ensure responsible and beneficial use of applied AI. Applied AI is the practical application of artificial intelligence technologies to tackle real-world challenges and optimize processes in industries and businesses. It focuses on deploying AI techniques and algorithms to achieve tangible outcomes and deliver value. Its significance lies in enhancing efficiency, productivity, and decision-making through AI-driven solutions, unlike theoretical AI research that focuses on advancing fundamental understanding (Patil & Shankar, 2023).

AI is playing a pivotal role in the competitive business landscape, enabling organizations to gain insights from vast data, automate repetitive tasks, and make data-driven decisions at scale. This transformative potential span various sectors, including manufacturing, finance, healthcare, and retail, demonstrating its transformative potential. AI can optimize processes and workflows within organizations, streamlining operations, reducing costs, and enhancing quality. Predictive maintenance algorithms can anticipate equipment failures, minimizing downtime and maximizing productivity in manufacturing plants. AI-powered supply chain management systems can optimize inventory levels and logistics routes, leading to efficient resource utilization and improved customer satisfaction (Aldoseri et al., 2023).

AI is revolutionizing traditional business functions like sales, marketing, and customer service by personalizing interactions, delivering targeted campaigns, and providing proactive support. Natural language processing algorithms analyze customer feedback, allowing businesses to tailor products and services to meet evolving needs. AI-powered chatbots and virtual assistants enhance customer satisfaction and loyalty. AI-driven solutions offer transformative potential beyond process optimization and customer engagement. They enable strategic decision-making and innovation by providing actionable insights from complex data sets. AI can optimize investment portfolios, predict market trends, and discover new drug candidates, empowering businesses to stay ahead of the curve and drive sustainable growth in a dynamic and uncertain environment (Ahmad, n.d.).

AI is transforming the way organizations use technology to achieve objectives and stay competitive in the digital age. It drives innovation, optimizes processes, and improves customer experiences across industries. As organizations adopt AI-driven solutions, its transformative potential will grow, creating a new era of opportunity and prosperity for businesses worldwide.

LEVERAGING AI FOR INDUSTRY OPTIMIZATION

AI technologies are revolutionizing various industries by automating tasks, improving decision-making, and driving efficiency gains. They enhance production, quality control, maintenance, logistics routes, inventory management, and supply chain visibility in manufacturing, and improve medical imaging analysis, personalized treatment plans, and administrative efficiency in healthcare. As AI continues to evolve, its impact on process optimization will continue to grow, leading to further advancements and efficiencies across these sectors(Zaki, 2019).

Figure 1. AI technologies optimize processes across different industries

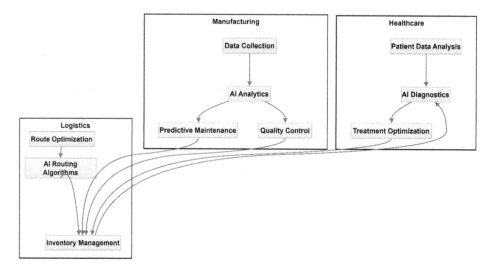

Figure 1 showcases how AI technologies, using data analytics, predictive algorithms, and automation, enhance efficiency, reduce costs, and improve outcomes across various industries.

Manufacturing

AI-powered predictive maintenance systems use sensor data to anticipate potential failures, reducing downtime and maintenance costs. Quality control is automated using computer vision algorithms to detect defects on assembly lines. Production optimization is achieved using real-time data, demand forecasts, and market trends to optimize schedules and resource allocation, enhancing efficiency, reducing waste, and maximizing output while minimizing costs (Naveeenkumar et al., 2024; Ravisankar et al., 2024).

Logistics

- Route Optimization: AI algorithms analyze various factors such as traffic patterns, weather conditions, and delivery constraints to optimize delivery routes. This results in faster deliveries, reduced fuel consumption, and lower operational costs for logistics companies.

- Inventory Management: AI-powered inventory management systems use predictive analytics to forecast demand, optimize stock levels, and prevent stockouts or overstocking. This minimizes inventory holding costs and ensures timely order fulfillment.
- Supply Chain Visibility: AI-driven supply chain visibility platforms provide real-time insights into the movement of goods across the supply chain. This enables companies to identify bottlenecks, mitigate risks, and make data-driven decisions to improve overall supply chain efficiency.

Healthcare (Pramila et al., 2023; Ramudu et al., 2023; Satav et al., 2023)

- Medical Imaging Analysis: AI algorithms analyze medical images such as X-rays, MRIs, and CT scans to assist radiologists in detecting abnormalities and diagnosing diseases. This improves diagnostic accuracy, reduces interpretation time, and enhances patient outcomes.
- Personalized Treatment Plans: AI-based predictive analytics leverage patient data and medical literature to predict treatment responses and recommend personalized treatment plans. This enables healthcare providers to deliver more effective and targeted interventions tailored to individual patient needs.
- Administrative Efficiency: AI-powered chatbots and virtual assistants automate administrative tasks such as appointment scheduling, billing inquiries, and patient triage. This frees up healthcare professionals' time, improves patient experience, and reduces administrative costs.

Real-world examples demonstrate the benefits of AI applications in driving efficiency, cost reduction, and quality improvement across industries, optimizing processes, enhancing decision-making, and delivering value to customers and stakeholders.

REVOLUTIONIZING BUSINESS OPERATIONS WITH AI

AI is revolutionizing sales, marketing, customer service, and human resources by enabling automation, personalization, and data-driven decision-making, transforming these areas through automation and data-driven strategies (Boopathi et al., 2023; Venkateswaran et al., 2023; Vennila et al., 2022).

Figure 2. AI is reshaping traditional business operations by automating tasks

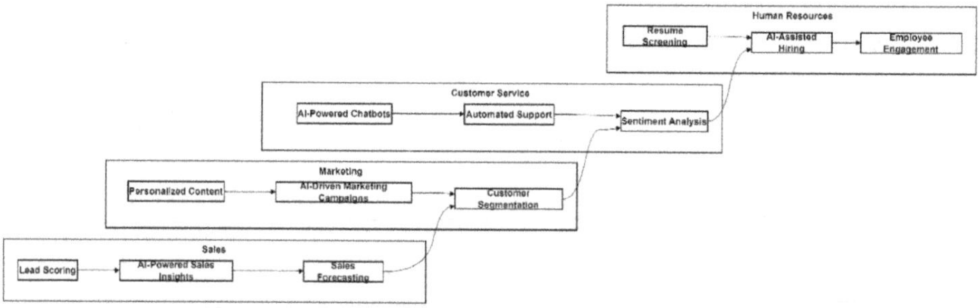

Figure 2 illustrates how AI is revolutionizing business operations by automating tasks, offering actionable insights, and personalizing experiences across sales, marketing, customer service, and human resources.

Sales

- **Lead Scoring and Prioritization:** AI algorithms analyze historical data and online behavior patterns to predict the likelihood of leads converting into customers. This enables sales teams to prioritize high-quality leads, optimize their outreach efforts, and focus on prospects with the highest potential for conversion.
- **Sales Forecasting:** AI-powered predictive analytics models forecast future sales trends based on historical data, market dynamics, and external factors. By providing accurate sales forecasts, AI helps businesses anticipate demand, optimize inventory levels, and allocate resources more effectively.
- **Sales Process Automation:** AI-powered sales automation tools streamline repetitive tasks such as data entry, email communications, and follow-up reminders. This frees up sales representatives' time, allowing them to focus on building relationships with prospects and closing deals more efficiently.

Marketing

- **Personalized Recommendations:** AI algorithms analyze customer data, preferences, and purchase history to deliver personalized product recommendations and marketing messages. This enhances customer engagement, increases conversion rates, and fosters brand loyalty.

- **Predictive Analytics:** AI-powered predictive analytics models forecast customer behavior, market trends, and campaign performance metrics. By identifying patterns and insights hidden within large datasets, marketers can optimize marketing strategies, allocate budgets more effectively, and improve ROI.
- **Content Generation:** AI-generated content tools use natural language processing (NLP) algorithms to create personalized marketing content, such as product descriptions, blog posts, and social media updates. This accelerates content production, enhances consistency, and ensures relevance to target audiences.

Customer Service

- **Chatbots and Virtual Assistants:** AI-powered chatbots and virtual assistants provide instant responses to customer inquiries, resolve common issues, and offer personalized recommendations. By automating routine interactions, businesses can deliver round-the-clock customer support, improve response times, and reduce support costs.
- **Sentiment Analysis:** AI algorithms analyze customer feedback across various channels, such as social media, emails, and surveys, to gauge sentiment and identify trends. This enables businesses to proactively address customer concerns, improve product offerings, and enhance overall customer satisfaction.
- **Call Center Optimization:** AI-driven speech recognition and natural language processing technologies analyze customer calls to identify keywords, sentiments, and trends. This helps businesses optimize call routing, improve agent performance, and enhance the quality of customer interactions.

Human Resources

- **Talent Acquisition:** AI-powered recruitment platforms use data analytics and machine learning algorithms to identify top candidates, match them with relevant job opportunities, and predict their likelihood of success. This streamlines the hiring process, reduces time-to-fill, and improves the quality of hires.
- **Employee Engagement:** AI-driven sentiment analysis tools monitor employee feedback, engagement surveys, and social media interactions to assess morale, identify areas for improvement, and proactively address issues.

This fosters a positive work environment, enhances employee satisfaction, and reduces turnover rates.

- **Training and Development:** AI-powered learning management systems (LMS) leverage adaptive learning algorithms to personalize training programs based on employees' skills, preferences, and performance. This enhances learning outcomes, accelerates skill development, and enables continuous upskilling and reskilling.

AI is revolutionizing business operations by automating tasks, personalizing interactions, and enabling data-driven decision-making across sales, marketing, customer service, and human resources functions. This improves efficiency, enhances customer experiences, and gives businesses a competitive edge. Case studies demonstrate how AI-powered tools and platforms enhance decision-making and productivity across various industries (Boopathi & Davim, 2023; Pramila et al., 2023; Ramudu et al., 2023).

- **Netflix - Personalized Content Recommendations:** Netflix utilizes AI algorithms to analyze user behavior, viewing history, and preferences to deliver personalized content recommendations. By leveraging machine learning techniques, Netflix tailors its vast library of movies and TV shows to individual users' tastes, increasing viewer engagement and retention. The recommendation engine accounts for factors such as genre preferences, viewing habits, and user ratings to suggest content that resonates with each subscriber. This personalized approach has significantly contributed to Netflix's success, with over 80% of content watched being driven by recommendations, leading to higher user satisfaction and longer subscription durations.
- **Amazon - Predictive Inventory Management:** Amazon employs AI-powered predictive analytics models to forecast customer demand and optimize inventory levels in its fulfillment centers. By analyzing historical sales data, seasonal trends, and external factors like weather patterns, Amazon can anticipate demand fluctuations and adjust inventory levels accordingly. This proactive approach to inventory management minimizes stockouts, reduces excess inventory holding costs, and ensures timely order fulfillment. As a result, Amazon maintains high customer satisfaction levels, improves operational efficiency, and maximizes profitability.
- **Salesforce - AI-Powered Sales Insights:** Salesforce integrates AI capabilities into its customer relationship management (CRM) platform to provide sales teams with actionable insights and predictive analytics. By analyzing sales data, customer interactions, and market trends, Salesforce's AI-powered

tools identify patterns, opportunities, and potential risks, empowering sales representatives to make informed decisions and prioritize their efforts effectively. For example, Einstein Analytics analyzes sales pipeline data to predict which deals are most likely to close, enabling sales teams to focus their resources on high-value opportunities. This enhances sales productivity, accelerates deal velocity, and drives revenue growth for Salesforce customers.

- **Google - Smart Email Responses with Gmail:** Google's Gmail incorporates AI-powered features, such as Smart Reply and Smart Compose, to enhance productivity and streamline email communication. Smart Reply suggests quick, contextually relevant responses based on the content of incoming emails, allowing users to respond promptly with minimal effort. Meanwhile, Smart Compose uses predictive text technology to autocomplete sentences and phrases as users type, reducing typing time and minimizing errors. These AI-driven features save users valuable time, increase email efficiency, and improve overall productivity in both personal and professional settings.

Case studies demonstrate how organizations use AI-powered tools and platforms to improve decision-making and productivity. These technologies enable businesses to make smarter decisions, optimize processes, and achieve greater efficiency. As AI evolves, its impact on decision-making and productivity will continue to grow, reshaping the future of work across various industries.

ENHANCING CUSTOMER EXPERIENCES THROUGH AI

AI enhances customer interactions, recommendations, and satisfaction by utilizing data-driven insights, predictive analytics, and automation, thereby enhancing overall customer experiences through important components (Revathi et al., 2024).

Figure 3. AI enhances customer satisfaction

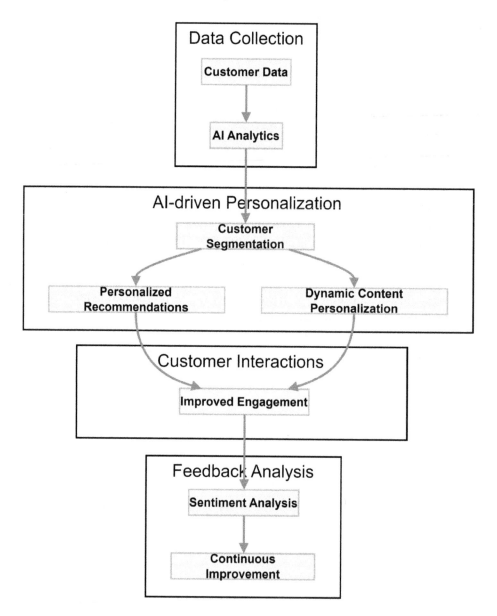

Figure 3 showcases how AI enhances customer satisfaction by personalizing interactions, providing tailored recommendations, and utilizing data analytics, segmentation, and dynamic personalization techniques.

Data-Driven Insights

- AI systems collect and analyze vast amounts of customer data from various sources, including purchase history, browsing behavior, demographics, and interactions with the brand.
- By processing this data, AI algorithms gain insights into individual preferences, interests, and needs, allowing businesses to better understand their customers on a granular level.
- These data-driven insights enable businesses to anticipate customer behavior, identify patterns, and tailor their interactions and offerings accordingly.

Personalized Recommendations

- AI-powered recommendation engines leverage machine learning algorithms to analyze customer data and predict products or services that are most relevant to each individual.
- These recommendation engines consider factors such as past purchases, browsing history, preferences, and similar users' behaviors to generate personalized recommendations in real-time.
- By presenting customers with personalized recommendations, businesses enhance the relevance of their offerings, increase cross-selling and upselling opportunities, and ultimately drive higher conversion rates and revenue.

Dynamic Content Personalization

- AI enables dynamic content personalization across various channels, including websites, emails, mobile apps, and social media platforms.
- Through A/B testing, user segmentation, and real-time optimization, AI systems deliver customized content and experiences tailored to each customer's preferences and behaviors.
- Dynamic content personalization enhances engagement, improves click-through rates, and fosters a deeper connection between the brand and the customer by delivering content that resonates with their interests and needs.

Automated Customer Service

- AI-powered chatbots and virtual assistants provide personalized customer support and assistance round-the-clock, addressing inquiries, resolving issues, and guiding customers through their purchase journey.
- Natural language processing (NLP) algorithms enable chatbots to understand and respond to customer queries in a conversational manner, simulating human-like interactions.
- By automating routine customer service tasks, AI-driven chatbots streamline the support process, reduce response times, and improve overall satisfaction by providing timely and accurate assistance.

Feedback Analysis and Sentiment Tracking

- AI algorithms analyze customer feedback, reviews, and social media mentions to gauge sentiment, identify trends, and gain insights into customer satisfaction levels.
- Sentiment analysis enables businesses to proactively address negative feedback, identify areas for improvement, and capitalize on positive sentiment to strengthen brand loyalty.
- By monitoring customer sentiment in real-time, businesses can make data-driven decisions to enhance the overall customer experience and drive continuous improvement.

AI is revolutionizing various industries by personalizing customer interactions, providing personalized recommendations, and enhancing satisfaction through data-driven insights, predictive analytics, and automation. This approach helps businesses build stronger relationships, increase loyalty, and achieve sustainable growth in a competitive market. AI-driven strategies enable businesses to stay competitive, meet evolving customer expectations, and drive sustainable growth, transforming sectors into innovative and efficient businesses (Naveeenkumar et al., 2024).

Retail

- **Virtual Styling and Personal Shopping:** Retailers like Sephora and Macy's utilize AI-powered virtual styling tools that analyze customers' preferences, skin tones, and style preferences to recommend personalized product selections. Customers can virtually try on makeup or outfits, receive personalized styling advice, and make informed purchasing decisions.

- **Predictive Inventory Management:** Retailers leverage AI algorithms to forecast demand, optimize inventory levels, and ensure product availability. By analyzing historical sales data, market trends, and external factors, AI-driven systems anticipate customer demand, minimize stockouts, and improve overall inventory management efficiency.

E-Commerce

- **Dynamic Pricing and Offer Optimization:** E-commerce platforms such as Amazon and eBay use AI-powered dynamic pricing algorithms to adjust prices in real-time based on factors like demand, competitor pricing, and customer behavior. This personalized pricing strategy maximizes revenue and enhances customer satisfaction by offering competitive prices and targeted discounts.
- **Personalized Product Recommendations:** E-commerce websites employ AI-driven recommendation engines to suggest relevant products to customers based on their browsing history, purchase patterns, and preferences. By delivering personalized recommendations, e-commerce platforms increase cross-selling opportunities, drive repeat purchases, and enhance customer engagement.

Hospitality and Travel

- **Chatbots for Booking and Customer Service:** Hospitality companies and travel agencies implement AI-powered chatbots to assist customers with booking accommodations, flights, and activities. Chatbots use natural language processing (NLP) to understand customer inquiries, provide personalized recommendations, and address common questions or concerns, improving the booking experience and enhancing customer satisfaction.
- **Dynamic Pricing and Room Allocation:** Hotels and airlines utilize AI algorithms to optimize pricing and room allocation based on factors like demand, seasonality, and historical booking patterns. By dynamically adjusting prices and room availability, hospitality businesses maximize revenue, optimize occupancy rates, and offer competitive rates to customers.

Banking and Finance

- **Personalized Financial Advice:** Banks and financial institutions leverage AI-powered robo-advisors to offer personalized investment advice and financial planning services to customers. Robo-advisors analyze customers' financial goals, risk tolerance, and investment preferences to recommend tailored investment portfolios and strategies, empowering customers to make informed decisions about their finances.
- **Fraud Detection and Prevention:** AI algorithms detect fraudulent activities and suspicious transactions in real-time by analyzing patterns, anomalies, and historical data. Banks use AI-driven fraud detection systems to identify potential threats, mitigate risks, and protect customers' accounts and assets, enhancing trust and confidence in banking services.

AI ETHICS AND GOVERNANCE

The widespread use of AI in industry and business necessitates careful attention and proactive measures to address ethical considerations (Boopathi & Khang, 2023).

Figure 4. Ethical AI development system

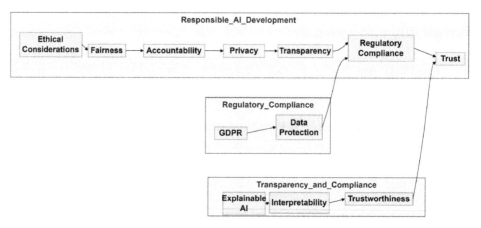

Figure 4 highlights the importance of transparency, regulatory compliance, and ethical AI development in creating trustworthy, fair, and trust-fostering systems.

Bias and Fairness

- AI algorithms may inadvertently perpetuate biases present in training data, leading to discriminatory outcomes in decision-making processes. For example, biased algorithms in hiring or lending processes could result in unfair treatment based on factors like race, gender, or socioeconomic status.
- To address this issue, organizations must ensure transparency and accountability in AI systems, regularly audit algorithms for bias, and implement measures to mitigate bias, such as diverse and representative training data sets and fairness-aware algorithms.

Privacy and Data Protection

- AI systems often rely on vast amounts of personal data to function effectively, raising concerns about privacy and data protection. Improper handling of sensitive information can lead to privacy breaches, identity theft, and unauthorized surveillance.
- Businesses must prioritize data privacy by implementing robust security measures, obtaining explicit consent for data collection and processing, and adhering to relevant regulations such as the General Data Protection Regulation (GDPR) and the California Consumer Privacy Act (CCPA).

Transparency and Accountability

- AI algorithms can be complex and opaque, making it difficult for users to understand how decisions are made or to challenge outcomes. Lack of transparency can erode trust and accountability in AI systems.
- Organizations should strive for transparency by documenting the development and deployment processes of AI systems, providing explanations for algorithmic decisions, and establishing mechanisms for recourse and redress in case of errors or biases.

Job Displacement and Economic Impact

- The widespread adoption of AI technologies may lead to job displacement and economic disruption, particularly in industries reliant on routine or repetitive tasks that can be automated.

- To mitigate the negative impact on employment, businesses should invest in reskilling and upskilling programs for affected workers, promote workforce diversity and inclusion, and explore opportunities for human-AI collaboration to augment rather than replace human labor.

Algorithmic Accountability

- AI systems are not infallible and may produce unintended consequences or errors that have real-world implications. Ensuring algorithmic accountability requires mechanisms for monitoring, auditing, and addressing the performance and impact of AI systems.
- Organizations should establish clear lines of responsibility and accountability for AI systems, conduct regular audits and assessments of algorithmic performance, and provide avenues for feedback and redress for individuals affected by algorithmic decisions.

Social Implications and Equity

- The deployment of AI technologies can have wide-ranging social implications, affecting issues such as access to essential services, distribution of resources, and societal norms.
- Businesses must consider the broader societal impact of their AI initiatives, engage with diverse stakeholders to understand their concerns and perspectives, and prioritize equity, inclusivity, and social responsibility in AI development and deployment.

The ethical use of AI requires a comprehensive approach that includes transparency, fairness, privacy, accountability, and social responsibility. By implementing ethical practices and engaging stakeholders, businesses can harness AI's transformative potential, mitigate risks, and foster trust in AI-driven systems (Boopathi & Khang, 2023). Responsible AI development, transparency, and regulatory compliance are crucial components.

Responsible AI development, transparency, and regulatory compliance are crucial for fostering trust, fairness, and ethical use of AI in society. By integrating these principles into AI development practices, organizations can create AI systems that promote human well-being, respect individual rights, and contribute to a more equitable future. These principles also offer strategic advantages, allowing organizations to differentiate themselves, mitigate risks, and build enduring relationships with customers, stakeholders, and communities.

Responsible AI Development

- Responsible AI development involves designing, building, and deploying AI systems in a manner that prioritizes ethical considerations, fairness, and societal well-being.
- It requires organizations to adhere to ethical principles such as transparency, accountability, fairness, privacy, and inclusivity throughout the AI development lifecycle.
- By adopting responsible AI practices, organizations can mitigate risks associated with bias, discrimination, and unintended consequences, while maximizing the benefits of AI for individuals, communities, and society as a whole.

Transparency

- Transparency in AI refers to the openness and clarity with which AI systems operate, enabling users to understand how decisions are made, why certain outcomes occur, and the underlying mechanisms driving AI behavior.
- Transparent AI systems foster trust, accountability, and user confidence by providing explanations for algorithmic decisions, disclosing sources of data and model biases, and enabling users to verify and validate AI-driven outcomes.
- Transparency also facilitates collaboration, peer review, and continuous improvement of AI systems, driving innovation and promoting responsible AI development practices across industries.

Regulatory Compliance

- Regulatory compliance ensures that AI systems adhere to applicable laws, regulations, and standards governing their development, deployment, and use.
- Regulatory frameworks such as the General Data Protection Regulation (GDPR), the California Consumer Privacy Act (CCPA), and sector-specific regulations (e.g., healthcare, finance) impose requirements related to data privacy, security, transparency, fairness, and accountability in AI applications.

- Compliance with regulatory requirements helps organizations mitigate legal and financial risks, protect individuals' rights and freedoms, and maintain public trust and confidence in AI technologies.

CHALLENGES AND FUTURE TRENDS

Implementing AI solutions presents challenges for organizations, including data privacy concerns, talent shortage, algorithmic biases, ethical compliance, and system integration. These issues require a holistic approach that includes technical expertise, ethical considerations, regulatory compliance, and organizational readiness. By proactively identifying and mitigating these challenges, organizations can unlock the full potential of AI technologies, drive innovation, efficiency, and growth in the digital age (Agrawal et al., 2023; Koshariya et al., 2023; Kumar et al., 2023; Samikannu et al., 2022).

Figure 5. Implementing AI solutions: Challenges and future trends

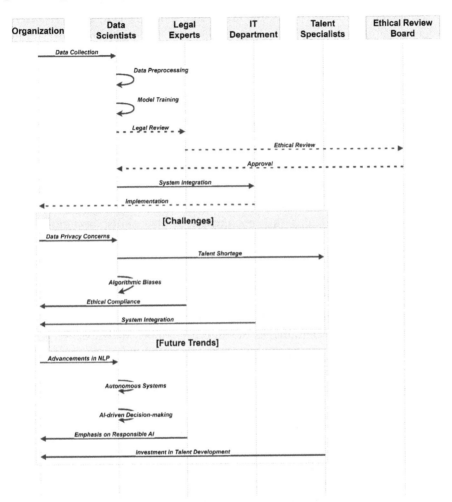

Figure 5 outlines the steps of implementing AI solutions, including data collection, preprocessing, model training, legal and ethical review, system integration, and implementation. It addresses challenges like data privacy, talent shortages, algorithmic biases, ethical compliance, and system integration, and future AI trends like NLP, autonomous systems, AI-driven decision-making, and talent development.

- **Data Privacy Concerns:** One of the primary challenges in implementing AI solutions revolves around data privacy concerns. AI algorithms require large volumes of data to train and operate effectively, often including sensitive information about individuals. Organizations must navigate complex regulatory frameworks, such as the GDPR and CCPA, to ensure compliance with data protection laws. Additionally, ensuring the security and confidentiality of data throughout its lifecycle, from collection and storage to processing and analysis, is paramount to safeguarding individuals' privacy rights. Failure to address data privacy concerns adequately can result in legal penalties, reputational damage, and loss of customer trust.

- **Talent Shortage:** Another significant challenge in implementing AI solutions is the shortage of skilled talent with expertise in AI technologies, machine learning, data science, and related fields. The demand for AI professionals far exceeds the supply, leading to fierce competition for top talent. Organizations face difficulty recruiting and retaining qualified AI professionals capable of designing, developing, and deploying AI solutions. Moreover, building multidisciplinary teams that encompass diverse skills, backgrounds, and perspectives is essential for addressing complex AI challenges effectively. Investing in talent development, training programs, and collaboration with educational institutions can help organizations bridge the talent gap and build internal AI capabilities.

- **Algorithmic Biases:** Algorithmic biases pose a significant challenge in implementing AI solutions, as AI algorithms may inadvertently perpetuate or amplify biases present in training data. Biased algorithms can lead to discriminatory outcomes in decision-making processes, exacerbating existing social inequalities and reinforcing systemic biases. Organizations must proactively identify and mitigate biases in AI systems by carefully curating training data, employing fairness-aware algorithms, and conducting rigorous testing and validation processes. Moreover, fostering diversity and inclusivity in AI development teams can help mitigate unconscious biases and promote fairness and equity in AI applications.

- **Ethical and Regulatory Compliance:** Ensuring ethical and regulatory compliance is a critical challenge in implementing AI solutions, as organizations must navigate complex ethical dilemmas and legal requirements governing AI development and deployment. Ethical considerations such as transparency, accountability, fairness, and privacy must be integrated into AI development practices to mitigate risks associated with bias, discrimination, and unintended consequences. Additionally, organizations must adhere to regulatory frameworks such as the GDPR, CCPA, and sector-specific regulations to protect individuals' rights, privacy, and autonomy. Failure to address ethical and regulatory concerns can lead to legal liabilities, regulatory fines, and damage to organizational reputation.

- **Integration with Existing Systems:** Integrating AI solutions with existing IT infrastructure and business processes presents technical challenges related to compatibility, scalability, and interoperability. Legacy systems may lack the flexibility and adaptability required to support AI implementations, requiring organizations to invest in modernization efforts and infrastructure upgrades. Moreover, ensuring seamless integration and interoperability between AI systems and other organizational systems, such as enterprise resource planning (ERP) systems, customer relationship management (CRM) platforms, and cybersecurity solutions, is essential for maximizing the value and impact of AI initiatives.

Future trends in applied AI include advancements in natural language processing, autonomous systems, and AI-driven decision-making, with potential trends shaping the future of applied AI (Boopathi, 2023; Ramudu et al., 2023; Revathi et al., 2024).

Natural Language Processing (NLP)

- **Conversational AI:** Future advancements in NLP will enable more natural and contextually aware interactions between humans and AI systems. Conversational AI technologies, including chatbots, virtual assistants, and voice-enabled interfaces, will become increasingly sophisticated, capable of understanding and responding to complex queries, emotions, and nuances in language.
- **Multilingual NLP:** As businesses operate in increasingly globalized markets, there will be a growing demand for multilingual NLP solutions capable of processing and understanding multiple languages. Advancements in machine translation, sentiment analysis, and language understanding will enable AI systems to communicate seamlessly across diverse linguistic and cultural contexts.

Autonomous Systems

- **Robotic Process Automation (RPA):** Autonomous systems powered by AI and machine learning algorithms will automate a wider range of tasks and processes across industries. RPA solutions will become more intelligent and adaptive, capable of autonomously performing complex tasks, learning from experience, and improving over time without human intervention.
- **Autonomous Vehicles:** The development and deployment of autonomous vehicles will continue to advance, driven by breakthroughs in sensor technology, computer vision, and reinforcement learning algorithms.

Autonomous vehicles will revolutionize transportation and logistics, offering safer, more efficient, and environmentally sustainable mobility solutions.

AI-Driven Decision-Making

- **Explainable AI (XAI):** There will be a growing emphasis on developing explainable AI (XAI) techniques that provide transparent and interpretable explanations for algorithmic decisions. XAI will enable users to understand how AI systems arrive at their conclusions, increasing trust, accountability, and regulatory compliance.
- **Augmented Intelligence:** AI-driven decision-making will shift towards a model of augmented intelligence, where AI systems complement human expertise rather than replacing it. Augmented intelligence tools will empower decision-makers with actionable insights, recommendations, and predictions, enabling more informed and effective decision-making across domains.

Ethical and Responsible AI

- **Ethical AI Frameworks:** There will be increasing focus on developing and implementing ethical AI frameworks and guidelines to ensure the responsible and ethical use of AI technologies. Organizations will prioritize transparency, fairness, accountability, and privacy in AI development and deployment practices, aligning with ethical principles and regulatory requirements.
- **Bias Mitigation and Fairness:** Efforts to mitigate algorithmic biases and ensure fairness in AI systems will intensify, driven by concerns about discrimination, inequity, and social impact. Advancements in fairness-aware AI algorithms and bias detection techniques will enable organizations to address biases systematically and proactively.

Future AI trends will focus on natural language processing, autonomous systems, and AI-driven decision-making, driven by research, technological innovation, and societal demand for intelligent and ethical solutions, shaping work, business, and innovation in industries, economies, and societies.

CONCLUSION

The chapter delves into the diverse applications of applied AI, highlighting its potential for transformation, ethical considerations, and future trends. AI has revolutionized industries by enabling automation, personalization, and data-driven decision-making, reshaping processes, enhancing efficiency, and driving innovation across various sectors. However, the widespread adoption of AI also presents challenges, including data privacy concerns, algorithmic biases, and talent shortages. Addressing these challenges requires a holistic approach that prioritizes responsible AI development, transparency, and regulatory compliance. By integrating ethical considerations, fairness, and accountability into AI development practices, organizations can mitigate risks, foster trust, and promote the ethical use of AI technologies.

Looking ahead, future trends in applied AI are expected to focus on advancements in natural language processing, autonomous systems, and AI-driven decision-making. Conversational AI, autonomous vehicles, and augmented intelligence tools will continue to evolve, offering new opportunities for innovation and growth. Moreover, ethical and responsible AI frameworks will play a crucial role in ensuring the ethical and equitable deployment of AI technologies, aligning with societal values and regulatory requirements. Applied AI is a transformative shift in business operations, innovation, and customer interaction. By addressing ethical challenges and embracing its transformative potential, organizations can drive sustainable growth and create positive societal impact in the digital age. Collaboration, transparency, and a commitment to ethical principles are crucial for maximizing AI's full potential for the benefit of all.

ABBREVIATIONS

- AI - Artificial Intelligence
- CCPA - California Consumer Privacy Act
- CRM - Customer Relationship Management
- CT - Computer Tomography
- ERP - Enterprise Resource Planning
- GDPR - General Data Protection Regulation
- IT - Information Technology
- LMS - Learning Management System
- MRI - Magnetic Resonance Imaging
- NLP - Natural Language Processing
- ROI - Return on Investment
- RPA - Robotic Process Automation

- TV - Television
- XAI - Explainable AI

REFERENCES

Agrawal, A. V., Shashibhushan, G., Pradeep, S., Padhi, S., Sugumar, D., & Boopathi, S. (2023). Synergizing Artificial Intelligence, 5G, and Cloud Computing for Efficient Energy Conversion Using Agricultural Waste. In Sustainable Science and Intelligent Technologies for Societal Development (pp. 475–497). IGI Global.

Ahmad, J. (n.d.). *Navigating the Future: Harnessing Artificial Intelligence for Business Success*. Academic Press.

Aldoseri, A., Al-Khalifa, K., & Hamouda, A. (2023). *A roadmap for integrating automation with process optimization for AI-powered digital transformation*. Academic Press.

Boopathi, S. (2023). Securing Healthcare Systems Integrated With IoT: Fundamentals, Applications, and Future Trends. In Dynamics of Swarm Intelligence Health Analysis for the Next Generation (pp. 186–209). IGI Global.

Boopathi, S., & Davim, J. P. (2023). *Sustainable Utilization of Nanoparticles and Nanofluids in Engineering Applications*. IGI Global. doi:10.4018/978-1-6684-9135-5

Boopathi, S., & Khang, A. (2023). AI-Integrated Technology for a Secure and Ethical Healthcare Ecosystem. In *AI and IoT-Based Technologies for Precision Medicine* (pp. 36–59). IGI Global. doi:10.4018/979-8-3693-0876-9.ch003

Boopathi, S., Kumar, P. K. S., Meena, R. S., Sudhakar, M., & Associates. (2023). Sustainable Developments of Modern Soil-Less Agro-Cultivation Systems: Aquaponic Culture. In Human Agro-Energy Optimization for Business and Industry (pp. 69–87). IGI Global.

Jarrahi, M. H., Kenyon, S., Brown, A., Donahue, C., & Wicher, C. (2023). Artificial intelligence: A strategy to harness its power through organizational learning. *The Journal of Business Strategy*, *44*(3), 126–135. doi:10.1108/JBS-11-2021-0182

Kitsios, F., & Kamariotou, M. (2021). Artificial intelligence and business strategy towards digital transformation: A research agenda. *Sustainability (Basel)*, *13*(4), 2025. doi:10.3390/su13042025

Koshariya, A. K., Kalaiyarasi, D., Jovith, A. A., Sivakami, T., Hasan, D. S., & Boopathi, S. (2023). AI-Enabled IoT and WSN-Integrated Smart Agriculture System. In *Artificial Intelligence Tools and Technologies for Smart Farming and Agriculture Practices* (pp. 200–218). IGI Global. doi:10.4018/978-1-6684-8516-3.ch011

Kumar, M., Kumar, K., Sasikala, P., Sampath, B., Gopi, B., & Sundaram, S. (2023). Sustainable Green Energy Generation From Waste Water: IoT and ML Integration. In Sustainable Science and Intelligent Technologies for Societal Development (pp. 440–463). IGI Global.

Naveeenkumar, N., Rallapalli, S., Sasikala, K., Priya, P. V., Husain, J., & Boopathi, S. (2024). Enhancing Consumer Behavior and Experience Through AI-Driven Insights Optimization. In *AI Impacts in Digital Consumer Behavior* (pp. 1–35). IGI Global. doi:10.4018/979-8-3693-1918-5.ch001

Patil, S., & Shankar, H. (2023). Transforming healthcare: Harnessing the power of AI in the modern era. *International Journal of Multidisciplinary Sciences and Arts*, 2(1), 60–70.

Pramila, P., Amudha, S., Saravanan, T., Sankar, S. R., Poongothai, E., & Boopathi, S. (2023). Design and Development of Robots for Medical Assistance: An Architectural Approach. In Contemporary Applications of Data Fusion for Advanced Healthcare Informatics (pp. 260–282). IGI Global.

Ramudu, K., Mohan, V. M., Jyothirmai, D., Prasad, D., Agrawal, R., & Boopathi, S. (2023). Machine Learning and Artificial Intelligence in Disease Prediction: Applications, Challenges, Limitations, Case Studies, and Future Directions. In Contemporary Applications of Data Fusion for Advanced Healthcare Informatics (pp. 297–318). IGI Global.

Ravisankar, A., Shanthi, A., Lavanya, S., Ramaratnam, M., Krishnamoorthy, V., & Boopathi, S. (2024). Harnessing 6G for Consumer-Centric Business Strategies Across Electronic Industries. In AI Impacts in Digital Consumer Behavior (pp. 241–270). IGI Global.

Revathi, S., Babu, M., Rajkumar, N., Meti, V. K. V., Kandavalli, S. R., & Boopathi, S. (2024). Unleashing the Future Potential of 4D Printing: Exploring Applications in Wearable Technology, Robotics, Energy, Transportation, and Fashion. In Human-Centered Approaches in Industry 5.0: Human-Machine Interaction, Virtual Reality Training, and Customer Sentiment Analysis (pp. 131–153). IGI Global.

Samikannu, R., Koshariya, A. K., Poornima, E., Ramesh, S., Kumar, A., & Boopathi, S. (2022). Sustainable Development in Modern Aquaponics Cultivation Systems Using IoT Technologies. In *Human Agro-Energy Optimization for Business and Industry* (pp. 105–127). IGI Global.

Satav, S. D., Hasan, D. S., Pitchai, R., Mohanaprakash, T., Sultanuddin, S., & Boopathi, S. (2023). Next generation of internet of things (ngiot) in healthcare systems. In *Sustainable Science and Intelligent Technologies for Societal Development* (pp. 307–330). IGI Global.

Suryadevara, C. K. (2023). Transforming Business Operations: Harnessing Artificial Intelligence and Machine Learning in the Enterprise. *International Journal of Creative Research Thoughts*, 2320–2882.

Venkateswaran, N., Vidhya, R., Naik, D. A., Raj, T. M., Munjal, N., & Boopathi, S. (2023). Study on Sentence and Question Formation Using Deep Learning Techniques. In *Digital Natives as a Disruptive Force in Asian Businesses and Societies* (pp. 252–273). IGI Global. doi:10.4018/978-1-6684-6782-4.ch015

Vennila, T., Karuna, M., Srivastava, B. K., Venugopal, J., Surakasi, R., & Sampath, B. (2022). New Strategies in Treatment and Enzymatic Processes: Ethanol Production From Sugarcane Bagasse. In Human Agro-Energy Optimization for Business and Industry (pp. 219–240). IGI Global.

Zaki, M. (2019). Digital transformation: Harnessing digital technologies for the next generation of services. *Journal of Services Marketing*, *33*(4), 429–435. doi:10.1108/JSM-01-2019-0034

Chapter 3
AI and Blockchain Revolution in Communication Education:
Future of Work in the Age of Industrial Robotics

C. V. Suresh Babu
https://orcid.org/0000-0002-8474-2882
Hindustan Institute of Technology and Science, India

Sudhir Manoharan
https://orcid.org/0009-0009-1326-7023
Hindustan Institute of Technology and Science, India

ABSTRACT

This chapter provides a comprehensive overview of AI and blockchain's role in communication education. It establishes the context, objectives, and significance of these technologies. The subsequent sections explore AI's impact in communication education, including content creation, personalization, and communication analysis. It also delves into blockchain's contributions, focusing on trust and transparency. The chapter addresses employment patterns, discussing automation's impact and the need for adaptability. It examines implications for professionals and educators, emphasizing ethical considerations and curriculum integration. Additionally, it explores enhancing pedagogy with AI and blockchain, offers strategic recommendations, showcases case studies, and looks ahead, emphasizing ethics and human-centered approaches.

DOI: 10.4018/979-8-3693-2399-1.ch003

1. INTRODUCTION

1.1 Background and Scope

The combination of AI and Blockchain in the area of the Future of Work represents a fundamental change in the way we interact with technology at work and in our daily lives (Suresh Babu C.V. 2022). The landscape of work changes as a result of AI automating repetitive jobs and Blockchain ensuring secure and decentralized transactions, necessitating a new skill set that emphasizes adaptability, innovation, and cooperation. The rise of new sectors and professional categories is prompted by this change, which also calls for a revaluation of ethical, privacy, and prejudice issues. A dynamic, skill-focused workforce will be fostered in the workplace of the future through harmonic human-machine collaboration that transcends regional boundaries.

1.2 Objective of Chapter

The goal of the chapter on "Future of Work in AI and Blockchain " is to investigate how the combination of these technologies is changing the nature of the workforce. It examines the implications of automation, changing skill sets, moral dilemmas, new employment positions, and cooperation, providing guidance for people, organizations, and governments on how to successfully navigate this rapidly changing environment.

1.3 Scope and Significance of AI and Blockchain in Communication Education

Integrating AI and Blockchain technology into communication education holds significant promise, offering a range of potential benefits. Personalized learning experiences, tailored to individual student needs, become achievable through AI algorithms that adapt to different learning styles. Language skills can be enhanced with AI-driven language processing tools, providing real-time feedback and fostering more effective communication (Sophia et al., 2023).

The inclusion of virtual and augmented reality (VR/AR) allows for the simulation of real-world communication scenarios, offering students immersive experiences to refine their practical skills. Automated evaluation, facilitated by AI, streamlines assessment processes, providing timely and constructive feedback to students.

Moreover, emphasizing ethical behavior is inherent in this strategy. AI can be programmed to instill ethical principles, preparing students for responsible communication practices in their future careers. This transformative approach aligns communication education with contemporary expectations, ensuring students acquire industry-relevant skills.

Blockchain's role in this context extends to secure credential verification, assuring the authenticity of students' achievements. It also facilitates data-driven insights, enabling educators to tailor curricula based on performance analytics (Suresh Babu, C. V. & Sanjai Das. 2023). Additionally, the integration of these technologies opens up possibilities for global cooperation, transcending geographical boundaries and preparing students for a connected, technology-driven professional landscape. Overall, the incorporation of AI and Blockchain enriches communication education, offering a forward-looking, skill-oriented, and globally relevant learning experience. (Elena et.al, 2023)

2. UNDERSTANDING THE IMPACT OF AI IN COMMUNICATION EDUCATION

2.1 AI Application in Content Creation and Curation

AI applications in content creation and curation have ushered in a transformative era by automating writing, design, and multimedia production while offering tailored content suggestions based on user preferences. This revolution extends across various sectors, benefitting marketers and researchers alike. Marketers leverage AI's prowess in content recommendations to refine their strategies, enhancing user experiences with personalized and relevant materials. The timely extraction of information from social media and news sources provides researchers with valuable insights, contributing to informed decision-making.(Sophia et al ..,2023)

In the contemporary data-driven environment, the integration of AI in content creation accelerates operations, ensuring efficiency and consistency. The automated processes foster innovation by freeing up human resources to focus on strategic and creative aspects, unburdened by routine tasks. This innovation, coupled with AI's ability to optimize content delivery, not only streamlines operations but also enhances the overall quality and relevance of the content presented to the audience (Suresh Babu C.V. 2022).

The symbiotic relationship between AI applications and content creation not only meets the demands of today's dynamic digital landscape but also propels industries toward a future characterized by efficiency, innovation, and enhanced user engagement.(Ben et al..,2023) As AI continues to evolve, its role in content creation and curation will likely deepen, providing businesses and researchers with powerful tools to navigate the complexities of information management and delivery in an increasingly interconnected and data-rich world.

2.2 AI-Driven Personalization and Audience Targeting

The landscape of content, suggestions, and marketing messages has undergone a paradigm shift with the integration of AI-driven customization and audience targeting, harnessing the potent capabilities of data analysis and machine learning (Sophia, 2023). By leveraging these technologies, businesses aim to finely tailor content to individual tastes, thereby enhancing engagement and conversion rates. The core strength lies in the ability to dissect audiences based on demographics and habits, providing nuanced insights that refine platform, product, and service strategies.

While the benefits are evident, the implementation of AI-driven customization raises critical questions surrounding bias and data privacy, demanding a nuanced and cautious approach. The categorization of audiences based on algorithms introduces concerns about potential biases in content delivery, emphasizing the importance of ethical considerations in the design and deployment of these systems. Moreover, the collection and utilization of personal data for customization purposes necessitate stringent controls to safeguard individual privacy and maintain user trust.

Despite these challenges, the strategic application of AI in customization and targeting has become integral to contemporary marketing practices. The nuanced understanding of audience preferences and behaviors allows businesses to craft more relevant and impactful messages, fostering a deeper connection with consumers. As technology continues to evolve, the responsible and ethical use of AI-driven customization will be pivotal in striking a balance between personalization and privacy, ensuring that businesses can effectively cater to individual tastes while respecting the boundaries of data protection and unbiased content delivery (Alexandre et al, 2023)

2.3 AI-Powered Communication Analysis and Insights

The integration of AI-powered communication analysis and insights represents a groundbreaking advancement in processing vast volumes of data through sophisticated algorithms, facilitating activities such as sentiment analysis, trend detection, and the evaluation of engagement metrics. This transformative technology has proven instrumental in revolutionizing the way businesses approach communication strategies and decision-making processes. By harnessing AI capabilities, organizations can efficiently analyze massive datasets, providing invaluable insights into public sentiment, identifying emerging trends, and quantifying engagement metrics with unparalleled precision.(Sophia et al..,2023)

Despite the undoubted benefits, the ethical considerations surrounding the deployment of AI in communication analysis remain paramount. The responsible use of such technology is crucial to ensure that privacy concerns, bias mitigation, and transparency are meticulously addressed. Organizations must navigate the delicate balance between extracting valuable insights and respecting ethical boundaries, recognizing that the misuse of AI-powered communication analysis could lead to unintended consequences.

Nevertheless, the potential applications of AI in this domain are vast. From refining communication tactics to assessing public opinion, organizations can leverage AI-powered insights to make well-informed decisions (Sophia et al..,2023) This technology enables a proactive and data-driven approach to communication strategies, allowing businesses to adapt swiftly to changing dynamics in the market and respond effectively to public sentiment. In essence, the integration of AI in communication analysis not only enhances the efficiency of decision-making processes but also empowers organizations to stay ahead of the curve in an increasingly dynamic and competitive landscape.

3. ROLE OF BLOCKCHAIN IN COMMUNICATION EDUCATION

3.1 Blockchain for Secure and Trustworthy Communication Channels

Blockchain technology, through the application of encryption and decentralized consensus mechanisms, establishes secure and trustworthy communication channels by creating an immutable record of transactions. The utilization of Blockchain's robust architecture provides a safeguard against unauthorized access and manipulation, ensuring data integrity, transparency, and immutability. This impenetrable ledger fundamentally transforms the way businesses handle communication, offering heightened levels of security and reliability.

The core strength of Blockchain lies in its ability to fortify data security, mitigating the risks associated with unauthorized tampering or access. By employing encryption techniques and a decentralized approach to consensus, it establishes a level of trust that traditional centralized systems often struggle to achieve. This not only enhances the overall integrity of data but also reduces the vulnerability to cyber threats and manipulation.

Beyond securing data, Blockchain technology holds the promise of revolutionizing various aspects of business operations. It enables encrypted messaging, ensuring the confidentiality of communications, and provides a robust framework for authenticating digital identities, thereby addressing critical concerns related to privacy and security. However, successful deployment of Blockchain technology requires tackling challenges such as scalability issues and ensuring compliance with regulatory frameworks.(Wang et al..,2023)

In essence, the adoption of Blockchain technology marks a pivotal shift in how businesses approach communication and data management. Its potential to secure information, facilitate encrypted messaging, and verify digital identities positions Blockchain as a transformative force in enhancing the security and reliability of communication channels, offering a decentralized and secure foundation for the future of digital interactions. Addressing scalability concerns and navigating regulatory landscapes will be essential to fully unlock the potential benefits of Blockchain in reshaping the landscape of secure and reliable communication.

3.2 Decentralization and Transparency in Media and Journalism

The integration of Blockchain technology has the potential to revolutionize the media and journalism landscape, introducing unprecedented decentralization and transparency. By utilizing Blockchain's distributed ledger, media content can be validated, timestamped, and securely stored, mitigating the risks of censorship and manipulation. Each alteration or addition to the content is meticulously recorded, ensuring an accurate and transparent history of modifications. This not only enhances the integrity of journalistic material but also fosters a level of transparency that counters misinformation and bolsters trust in the media.

Moreover, Blockchain's capabilities extend to redefining revenue paradigms in media. The implementation of Blockchain -enabled micropayments and smart contracts opens avenues for direct interactions between journalists and their audiences, creating a more direct and sustainable financial relationship. This not only empowers content creators, including journalists and artists, but also establishes a more equitable distribution of revenue within the media ecosystem. (Ben et al., 2023).

While the potential benefits are promising, the widespread adoption of Blockchain in media and journalism faces challenges. Overcoming scalability issues, gaining user acceptance, and navigating complex legal frameworks are imperative for the full realization of decentralization and transparency in this domain. Tackling these obstacles will be essential to unlock the transformative potential of Blockchain, ensuring that media and journalism evolve into more decentralized, transparent, and financially sustainable entities that better serve both content creators and consumers in the digital age.

3.3 Blockchain-Based Credentialing and Certification in Communication

The integration of Blockchain technology in credentialing and certification within the field of communication marks a transformative paradigm shift, offering a secure, impenetrable, and decentralized platform for validating and preserving professional achievements. Through Blockchain's cryptographic methods and distributed ledger, individuals can securely store credentials, including degrees, certificates, and communicative accomplishments, in a transparent and immutable manner. This not only enhances the legitimacy and reliability of qualifications but also streamlines verification processes, reducing the burden of paperwork and mitigating the risk of fraud, as evidenced by research by Ze Wang et al. (2023).

One of the key advantages of Blockchain -based credentialing is that it empowers individuals to take ownership of their data, allowing for selective distribution of credentials. This gives professionals greater control over the dissemination of their achievements, promoting transparency and privacy simultaneously. However, for these Blockchain -based credentialing systems to gain widespread acceptance in the communication sector, they must overcome challenges related to standardization, interoperability, and the development of user-friendly interfaces. Achieving seamless integration and usability is crucial to ensuring that Blockchain -based credentialing becomes an accessible and widely adopted solution within the communication industry, providing a trustworthy and efficient means of verifying professional accomplishments and qualifications. As these obstacles are addressed, the potential impact of Blockchain on credentialing in communication holds promise for a future where verification processes are not only more secure and reliable but also more user-centric and streamlined.

4. TRANSFORMING EMPLOYMENT PATTERNS IN COMMUNICATION INDUSTRY

4.1 Automation and AI's Influence on Traditional Communication Rules

Blockchain -based credentialing and certification in communication represent a revolutionary paradigm shift, providing a secure, tamper-proof, and decentralized platform for validating and archiving professional achievements (Suresh Babu, C. V. & Padma. R., 2023). Leveraging the cryptographic methods and distributed ledger of Blockchain technology, individuals can securely store credentials like degrees, certificates, and communicative successes in a transparent and irreversible manner. This not only elevates the legitimacy and reliability of certifications but also streamlines verification processes, effectively reducing administrative costs and mitigating the risks of fraud—a transformative impact. (Yang Liu et al. 2023).

The decentralized nature of Blockchain -based credentialing empowers individuals to take control of their data, enabling selective distribution of credentials. This not only enhances privacy but also ensures professionals have greater autonomy over how their achievements are shared and verified. Nevertheless, for these credentialing systems to gain widespread adoption within the communication industry, hurdles related to standardization, interoperability, and the development of user-friendly interfaces must be addressed(Le Wang et al, 2023).Overcoming these challenges is pivotal to ensuring that Blockchain -based credentialing becomes a practical and accessible solution in communication, where the verification of professional accomplishments is not only more secure and reliable but also more user-centric and cost-effective. As these barriers are gradually dismantled, the potential of Blockchain in credentialing promises a future where the validation of professional achievements is not only technologically advanced but also universally embraced across the communication sector.

4.2 New Job Opportunities Arising From AI and Blockchain Integration

The integration of Blockchain and AI is reshaping the employment landscape, fostering a demand for a diverse skill set across various sectors. Professionals proficient in artificial intelligence, machine learning, data science, and Blockchain development are increasingly sought after to build and manage these technologies. The role of Blockchain architects becomes crucial in ensuring secure solutions, while ethical AI specialists play a pivotal role in guaranteeing responsible and mindful implementation. AI instructors contribute to refining algorithms, and the expertise of Blockchain implementation specialists becomes essential in providing guidance throughout the integration process. This dynamic intersection of AI and Blockchain is giving rise to a spectrum of new and specialized career possibilities. As these technologies continue to evolve, there will be an emergence of novel job roles that require not only a deep understanding of AI and Blockchain but also a multidisciplinary approach to address the evolving demands of integration. The demand for professionals capable of navigating the intricacies of both technologies underscores the need for flexible knowledge and a versatile skill set to effectively harness the transformative potential of AI and Blockchain integration across diverse sectors. This evolving job landscape reflects the ongoing synergy between AI and Blockchain, creating a workforce that can adapt to the changing technological landscape and contribute to the responsible and innovative development of these powerful technologies.

4.3 Addressing Workforce Adaptability and Upskilling Needs

In the rapidly evolving landscape shaped by AI and Blockchain, addressing the imperatives of workforce flexibility and upskilling emerges as a critical mandate.

Continuous learning opportunities are paramount to ensuring that employees can acquire the dynamic skill sets demanded by these transformative technologies. Companies play a pivotal role in this paradigm by offering comprehensive training programs encompassing Blockchain, AI, and related subjects, emphasizing not just technical competencies but also fostering critical thinking, teamwork, and problem-solving abilities. As individuals must embrace the ethos of lifelong learning to remain competitive and adaptable amid technological shifts, a collaborative effort among governments, educational institutions, and corporations is imperative. This collaboration should be directed toward providing accessible and inclusive reskilling initiatives that equip the workforce with the expertise needed in this evolving landscape. By fostering a culture of continuous learning and collaboration, stakeholders can ensure the development of a highly trained and adaptive workforce capable of thriving in the workplace of the future. This collective commitment to ongoing education and skill enhancement stands as a cornerstone for navigating the challenges and harnessing the opportunities presented by the dynamic convergence of AI and Blockchain technologies.

5. IMPLICATIONS FOR COMMUNICATION PROFESSIONALS AND EDUCATORS

5.1 Navigating Ethical Considerations in AI-Driven Communication

Navigating ethical considerations in AI-driven communication is imperative to ensure ethical and impartial practices. Transparent communication about the role and potential impacts of AI is foundational, fostering understanding and trust. Addressing algorithmic bias is critical to prevent discriminatory outcomes, demanding ongoing monitoring and mitigation efforts to rectify unintended repercussions. Upholding principles of consent and secure data storage becomes paramount to safeguard privacy in the AI-driven landscape (Aydin bal et al.,2023). Vigilance against manipulation and misinformation is crucial when AI interacts with audiences, preserving the integrity of information dissemination.

Achieving a balance between innovation and ethical integrity necessitates collaboration between engineers, communication experts, and ethicists. This interdisciplinary approach is essential in crafting ethical guidelines and rules that govern AI-driven communication practices. By bringing together diverse perspectives, these collaborative efforts strive to establish a framework that not only encourages innovation but also upholds ethical standards. Such collaboration not only helps in addressing ethical challenges but also promotes credibility and confidence in

AI-driven communication system(Aydin et al..,2023).The synergy of technical expertise, communication insights, and ethical considerations contributes to the development of responsible AI practices that align with societal values, ensuring that these technologies contribute positively to the communication landscape while upholding ethical standards and respecting the rights of individuals.

5.2 Challenges in Integrating Blockchain and AI in Communication Curricula

Successfully integrating AI and Blockchain into communication courses poses several challenges. The rapid evolution of these technologies demands constant curriculum updates, necessitating a dynamic approach to stay abreast of advancements. Locating proficient instructors with expertise in both communication and AI/Blockchain proves challenging, highlighting the need for interdisciplinary collaboration and teacher training initiatives. Technological complexities may overwhelm students unfamiliar with these fields, exacerbated by limited access to resources, particularly in budget-constrained institutions (Ze Wang et al., 2023). Ethical considerations such as algorithmic bias and data privacy must be integral to the program, emphasizing the importance of ethical education. Striking the right balance between classroom instruction and real-world application is crucial, requiring practical projects and industry partnerships.

To surmount these obstacles, a collaborative effort involving academia, business, and policymakers is essential. This collaboration can foster innovation in communication education by creating comprehensive, flexible, and forward-looking courses. Regular dialogue with industry stakeholders can inform curriculum development, ensuring its relevance to real-world needs (Alexandre et al., 2023). Encouraging partnerships with technology providers can address resource constraints, while ethical discussions involving policymakers can shape responsible AI and Blockchain education. Establishing advisory boards comprising professionals from diverse fields can offer valuable insights, contributing to a well-rounded educational experience. By navigating these challenges collectively, stakeholders can create an educational framework that prepares students for the dynamic intersection of communication, AI, and Blockchain technologies, fostering a generation of professionals equipped to navigate the evolving landscape with ethical responsibility and practical competence (Aydin bal et al., 2023).

5.3 Fostering Innovation and Collaboration in Communication Education

Successfully integrating AI and Blockchain into communication courses presents a myriad of challenges. The rapid evolution of these technologies necessitates constant curriculum updates, demanding agility and responsiveness from educational institutions. Locating qualified instructors proficient in both communication and AI/Blockchain poses a formidable challenge, highlighting the need for specialized expertise (Suresh Babu, C. V., and B. Rohan. 2003). Technical complexities may overwhelm students unfamiliar with these fields, particularly considering potential resource constraints, especially for institutions with tight budgets.

Ensuring equitable access to resources and technology becomes a critical consideration, underscoring the importance of addressing disparities in educational infrastructure. Ethical concerns such as algorithmic bias and data privacy must be woven into the curriculum to foster responsible use of AI and Blockchain technologies. Striking a delicate balance between theoretical classroom instruction and real-world applications is imperative to bridge the gap between academic knowledge and practical implementation.

Overcoming these obstacles demands a collaborative effort between academia, businesses, and policymakers. Creating comprehensive, flexible, and forward-looking communication courses requires a concerted initiative to establish robust partnerships. By fostering collaboration, these stakeholders can collectively design educational programs that not only keep pace with technological advancements but also address the ethical implications and practical challenges associated with AI and Blockchain integration. The collaboration between academic institutions and industry experts ensures that curricula are not only relevant but also prepare students for the dynamic landscape of AI and Blockchain in the professional sphere. This collaborative approach is integral to shaping the future of communication education, aligning academic content with industry needs and fostering a generation of professionals adept at navigating the complexities of these cutting-edge technologies. (Sophia et al..,2023).

6. ENHANCING COMMUNICATION PEDAGOGY WITH AI AND BLOCKCHAIN

6.1 Integrating AI Tools In Communication Classroom Setting

The integration of AI tools into communication classroom settings presents both exciting prospects and notable challenges. The potential for AI to offer individualized feedback, enhance language abilities through language processing technologies, and improve interactive learning experiences is promising (Suresh Babu C.V. 2022). This technological advancement can contribute to a more personalized and effective educational environment. However, to harness these benefits, it is imperative to provide educators with adequate training to effectively utilize AI technologies and interpret their outcomes.

While AI has the potential to enrich learning experiences, there are concerns that need careful planning and consideration. Worries about AI potentially hindering creativity and critical thinking skills underscore the importance of thoughtful implementation strategies.(Ze Wang et al.., 2023). Balancing the integration of AI with the preservation of a friendly learning atmosphere is crucial. Ensuring fair access to technology is also a priority, especially in diverse educational settings where disparities in resource availability may exist.

Maintaining a well-rounded education requires the coexistence of traditional teaching approaches alongside AI integration. Striking the right balance is essential to avoid over-reliance on technology and to foster a holistic development of students. By strategically deploying AI tools and continually assessing their impact, educators can design engaging and enhanced communication learning experiences. This approach aims to prepare students for the evolving digital landscape while preserving the core values of critical thinking, creativity, and a supportive learning environment. Overall, the successful integration of AI in communication classrooms requires a thoughtful and collaborative effort, emphasizing training, accessibility, and a balanced educational approach.

6.2 Utilising Blockchain for Student Project and Collaborations

The integration of Blockchain technology into educational institutions has the potential to revolutionize transparency, authenticity, and teamwork, particularly in student projects and partnerships. Blockchain ensures accountability and prevents unauthorized modifications by establishing a secure and immutable record of project progress, contributions, and interactions. The use of smart contracts can automate aspects of collaboration, such as work assignments and incentive distribution, further enhancing efficiency and trust (Ben et al., 2023).

One significant advantage of Blockchain in education is its ability to facilitate inter-institutional partnerships. By securely exchanging data while preserving individual privacy, Blockchain opens avenues for seamless collaboration between educational entities. This not only fosters a more interconnected educational ecosystem but also ensures the integrity and security of shared information.

However, the adoption of Blockchain in education is not without its challenges. Overcoming obstacles such as educating teaching staff and students about Blockchain, integrating it into existing systems, and addressing potential scalability issues is crucial for successful implementation. Despite these challenges, the benefits of Blockchain in education are compelling (Sophia et al..,2023). By instilling a higher degree of transparency and trust in student projects and teamwork, Blockchain prepares students for technologically advanced professional environments. It offers a glimpse into a future where educational interactions are characterized by enhanced security, efficiency, and collaboration, laying the foundation for a more advanced and interconnected learning landscape.

6.3 Leveraging AI and Blockchain for Research and Publication

The convergence of Blockchain and AI holds transformative potential for academic publishing and research practices. AI emerges as a powerful ally in this context, enabling academics to analyze vast datasets swiftly, identify trends, and derive data-driven insights. It streamlines literature reviews, automates tedious tasks, and elevates the overall quality of research by expediting processes that would traditionally be time-consuming. Blockchain technology contributes to the revolution by ensuring immutability and transparency throughout the publishing process. By creating an unalterable record, Blockchain prevents data tampering, fostering confidence in the authenticity of research findings.

Smart contracts, facilitated by Blockchain, offer the possibility of automating administrative tasks such as copyright management and peer review processes. This not only enhances efficiency but also reduces the potential for human error and bias. However, challenges include the need to address biases in AI-generated content and seamlessly integrate these advanced technologies into existing workflows.(Vasco et al..,2023).Overcoming these hurdles is crucial for realizing the full potential of AI and Blockchain in reshaping academic publishing.

The synergy between AI and Blockchain has the capacity to revolutionize the entire research landscape. Accelerating the research process, fostering collaboration, and maintaining the integrity of academic publications are among the transformative outcomes. By leveraging the capabilities of AI for data analysis and Blockchain for secure, transparent record-keeping, the academic community can embrace a future where information is produced, shared, and validated with unprecedented speed, efficiency, and reliability.

7. STRATEGIC RECOMMENDATION FOR COMMUNICATION EDUCATION

7.1 Adapting Curricula to Embrace AI and Blockchain in Communication

Adapting communication curricula to incorporate the principles of Blockchain and AI is imperative for preparing students for the evolving workplace landscape. A fundamental understanding of AI and Blockchain embedded in the curriculum equips graduates with the knowledge to comprehend and leverage their applications effectively. To enhance job readiness, it is essential to integrate practical skills into the curriculum, enabling students to apply AI-driven analytics for audience insights or employ Blockchain for secure communication. The promotion of interdisciplinary approaches, intertwining ethics, technology, and communication, contributes to a well-rounded knowledge base.

To address the rapid evolution of these technologies, regular updates to the curriculum are crucial. The gap between academic knowledge and real-world application is effectively bridged by fostering collaboration with industry professionals and incorporating immersive learning opportunities (Ben et al., 2023). Real-world projects utilizing AI and Blockchain not only provide hands-on experience but also ensure that students are well-versed in applying theoretical concepts to practical scenarios.

In essence, the modification of curricula serves as a strategic response to the demands of the contemporary communication landscape. Working in tandem with business leaders and integrating real-world applications ensures that graduates are not only acquainted with the latest technological advancements but are also adept at navigating and leveraging the power of AI and Blockchain in their professional endeavors (Ben et al., 2023). By embracing these changes in education, students are empowered to traverse the complex terrain of modern communication with efficiency and innovation.

7.2 Facilitating Industry-Academia Collaboration for Skill Development

Closing the gap between academic curriculum and industry demands necessitates the promotion of robust industry-academia partnerships for talent development. Through strategic collaborations, educational institutions can align their curricula with the dynamic requirements of the job market. Regular workshops, seminars, and guest lectures conducted by industry professionals provide students with insights into current industry practices, fostering a more practical understanding of their field. The integration of collaborative projects and internship opportunities further exposes students to real-world challenges, enhancing their ability to apply theoretical knowledge in practical scenarios (Vasco et al.,2023)

Industry participation in curriculum development ensures the inclusion of current and relevant material, keeping educational content in sync with the rapidly evolving demands of the professional landscape. However, challenges such as maintaining a balance between theoretical knowledge and practical skills, as well as coordinating academic deadlines with business expectations, need to be effectively addressed for these partnerships to thrive.

Ultimately, these collaborations create a symbiotic relationship wherein graduates are better equipped for the workforce, possessing skills directly aligned with industry needs. Simultaneously, businesses gain access to a pool of competent professionals who can readily meet their evolving requirements (Vasco et al..,2023). This mutually beneficial connection not only enhances the employability of graduates but also contributes to the adaptability of businesses by providing them with a workforce that is well-prepared and responsive to the ever-changing dynamics of the industry.

7.3 Promoting Diversity and Inclusion in AI-Driven Communication

The ethical and objective deployment of AI-driven communication hinges on a proactive commitment to promoting inclusion and diversity. Mitigating biases in algorithms, which could perpetuate discriminatory behaviors, becomes more attainable by ensuring diverse representation within AI development teams. Fundamental to this approach is the imperative to train AI systems using diverse datasets and regularly scrutinize their performance for potential bias. The incorporation of inclusive design principles into AI-driven interfaces serves as a crucial step in ensuring universal accessibility. (Elena et al..,2023)

To bolster awareness and sensitivity, it is essential to encompass a wide array of use cases in AI systems and foster open discussions on ethical considerations. This approach not only educates stakeholders but also encourages a collective approach to identifying and rectifying potential biases (Sophia et al..,2023) In the pursuit of more inclusive technology, collaborative partnerships between communicators, communication experts, and marginalized populations can be transformative. By actively involving those who represent diverse perspectives, these partnerships contribute to the creation of technology that is inherently more inclusive and considerate of varied user needs.

Ultimately, the emphasis on diversity and inclusiveness in AI-driven communication translates into fair and morally sound outcomes, which serve the broader interests of society. This approach not only aligns with ethical imperatives but also acknowledges the societal impact of technology(Elena et al..,2023) .By prioritizing inclusivity, the trajectory of AI-driven communication becomes one that embraces diversity, ensuring that technological advancements benefit everyone equitably and contribute to a more just and harmonious social fabric.

8. CASE STUDIES: AI AND BLOCKCHAIN INTEGRATION IN COMMUNICATION EDUCATION

8.1 Successful Examples of AI-Driven Communication Courses

The potential for using AI into education has been demonstrated by the emergence of several successful instances of AI-driven communication courses

i. **Using Natural Language Processing** to Create Content courses that show students how to create high-quality written material using AI-powered technologies. Students get knowledge about how to use AI algorithms for effective content creation.

ii. **Social Media Insights and Analytics** courses that emphasize the use of AI-driven technologies to examine social media data in order to get audience insights, assess sentiment, and spot trends. When it comes to social media marketing, students learn to make judgments based on facts.

iii. Courses that examine the use of **AI in journalism**, such as data-driven storytelling, automated news writing, and investigative reporting utilizing AI-powered technologies.

iv. Courses that explore the **ethical issues of artificial intelligence and communication**

8.2 Blockchain-Powered Journalism and Media Initiative

Initiatives in journalism and media driven by Blockchain have shown how this technology may improve media's transparency, credibility, and sustainability

i. **Decentralized content verification** is the first step. By timestamping and preserving material on an immutable ledger, Blockchain protects the veracity of news. As a result, readers are more likely to believe what they read.

ii. **Monetization and Micropayments**, Blockchain transforms monetization strategies and lessens dependency on middlemen by enabling direct micropayments from readers to content authors.

iii. **Work Attribution and Ownership,** Blockchain enables transparent ownership records and explicit attribution of work to its producers, protecting against plagiarism and legal wrangling.

iv. **Freedom of expression and opposition to censorship** by preventing centralized control, decentralization makes it more difficult for governments or other groups to filter or repress information.

v. Blockchain can **enable safe**, decentralized cooperation in crowdsourced journalism.

8.3 Lesson Learned and Best Practices

The need of flexible curriculum that handle quick technical advancements, ethical issues, hands-on practicality, and multidisciplinary cooperation to imitate industrial dynamics is highlighted by lessons learnt through AI-driven communication courses. The best strategies include working with the industry, updating the curriculum to reflect the most recent AI developments, developing a wide range of skills, giving students real-world tasks, and promoting lifelong learning. (Signe Agerskov et al, 2023).

In the realm of Blockchain -powered journalism and media activities, pivotal lessons underscore the importance of prioritizing user-friendly interfaces, regulatory comprehension, and stakeholder education. To optimize the potential benefits, adherence to best practices becomes imperative.

Blockchain's integration can enhance transparency within media by providing an immutable ledger for content transactions.(Signe Agerskov et al,…2023). Additionally, leveraging Blockchain facilitates inclusive monetization through micropayments, ensuring fair compensation for content creators. Accurate content attribution is another significant advantage, bolstering credibility and combating misinformation.

The decentralization Inherent in Blockchain becomes a powerful tool against censorship, empowering journalists and content creators to resist undue influence (Signe Agerskov et al,023). Successful implementation, however, demands collaboration with Blockchain specialists who can navigate the technical intricacies and optimize the technology's potential.

User-friendly interfaces are crucial to democratize access, ensuring that journalists and users alike can easily engage with Blockchain -powered platforms. Regulatory understanding is equally vital, as compliance fosters legitimacy and trust within the industry.

In summary, embracing best practices in Blockchain -powered journalism involves transparent transactions, inclusive monetization, accurate attribution, resistance to censorship, and collaborative efforts with specialists (Alexandre et al., 2023). This approach, coupled with user-friendly design and regulatory awareness, forms the foundation for a transformative and ethically sound integration of Blockchain in media activities.

9. LOOKING AHEAD THE FUTURE OF COMMUNICATION EDUCATION

9.1 Anticipating Further Disruptions in the Age of Industrial Robotics

A comprehensive understanding of the transformative impacts and potential challenges associated with advanced automation technologies is imperative for anticipating future disruptions in the era of industrial robots. Diverse industries, including manufacturing, logistics, and healthcare, are poised to undergo substantial changes in worker demographics, skill requirements, production efficiency, and business models as industrial robots continue to advance. While the promises of higher productivity, enhanced accuracy, and cost-effectiveness accompany automation, concerns about job displacement and the imperative for upskilling and reskilling to align with evolving employment responsibilities loom large.(Alexandre et al..,2023).

The advent of industrial robots also introduces the prospect of human-machine collaborations and collaborative robots, altering traditional workplace dynamics. This shift necessitates the development of novel approaches to task assignment and staff training, recognizing the symbiotic relationship between human workers and automated technologies (Alexandre et al..,2023). Addressing challenges related to cybersecurity and ethics in AI-driven systems becomes paramount for ensuring safe and sustainable operations in this increasingly automated landscape. Striking a delicate balance between the benefits of automation and the potential socio-economic impacts is crucial for charting a course that maximizes the advantages of advanced technologies while mitigating potential disruptions and fostering a harmonious coexistence between human workers and their robotic counterparts. In navigating this transformative landscape, industries must remain adaptive, proactive, and ethically conscious to harness the full potential of industrial robots for the betterment of society and the workforce.

9.2 Ethical Guidelines and Policy Framework for AI and Blockchain in Communication

To ensure the ethical and responsible integration of emerging technologies like AI and Blockchain in communication, the establishment of robust ethical standards and regulatory frameworks is imperative. Ethical considerations encompass addressing biases embedded in AI systems, safeguarding data privacy, and promoting transparency in AI-driven decision-making processes. Bias mitigation is crucial to prevent discriminatory outcomes, while stringent data privacy measures are essential to protect sensitive information.

Blockchain's decentralized nature offers potential improvements in data security and transparency. However, safeguards must be implemented to prevent misuse and ensure compliance with data protection laws (Sophia et..,2023). Striking a balance between the advantages of decentralization and the necessity for regulatory control is vital. This involves developing frameworks that foster openness in technology deployment, enabling stakeholders to understand and trust the systems in place.

In summary, navigating the ethical terrain involves mitigating biases, safeguarding privacy, and fostering transparency. The decentralized nature of Blockchain introduces promising security and transparency benefits, but a careful regulatory approach is essential to prevent misuse and uphold data protection standards. Together, these measures form the foundation for a responsible and ethical integration of AI and Blockchain technologies in communication. (Signe Agerskov et al, 2023)

Principles for data ownership, permission, and secure sharing in both AI and Blockchain applications should be outlined in policy frameworks. To develop comprehensive recommendations, cross-sector collaboration including technological specialists, communicators, ethicists, and legislators is necessary (Signe Agerskov et al, 2023). These frameworks must be updated frequently to keep up with the rapid advancement of technology and new ethical dilemmas. In the end, well established ethical standards and regulatory frameworks.

9.3 Embracing a Human-Centered Approach in AI and Blockchain Integration

To make sure technology meets human wants and values, AI and Blockchain integration must adopt a human-centered approach. This requires building AI systems with user safety, inclusiveness, and well-being as top priorities. Key principles include avoiding biases, promoting transparency, and giving user control over AI interactions. Similar to this, a human-centered approach in Blockchain emphasizes user-friendly interfaces, data privacy, and decentralization that empowers individuals.

Effective development of AI and Blockchain applications hinges on prioritizing user input and feedback. By incorporating the perspectives of end users, technologists, social scientists, and ethicists, a holistic approach is achieved, ensuring that practical problems are addressed and user experiences are continually enhanced,(Ben Shneiderman et al..,2023), recognizes the importance of diverse expertise in creating technologies that align with societal needs and values.

Furthermore, the potential societal impact of AI and Blockchain extends beyond mere technical advancements. When designed with a human-centric focus, these technologies have the capacity to contribute significantly to social progress, uphold individual rights, and enhance overall human well-being. Placing humans at the center of technical breakthroughs emphasizes the ethical dimension of development, steering the trajectory of innovation towards positive societal outcomes.

In essence, fostering collaboration among various stakeholders and emphasizing the human perspective in AI and Blockchain development ensures not only functional and user-friendly applications but also positions technology as a force for positive social change and individual empowerment.

10. CONCLUSION

10.1 Recapitulation of Key Finding

Key discoveries that influence the effect of AI and Blockchain technology across numerous sectors are produced through their combination. Automation, personalization, and data-driven insights are improved by AI, while data security, transparency, and immutability are ensured by Blockchain. The changing nature of the workforce necessitates skill adaptability and ethical issues. The importance of collaboration, diversity, and multidisciplinary methods cannot be overstated, yet there are still issues with scalability and compliance. Well-being and ethics are given priority in a human-centered approach. A trust-based strategy encourages responsible adoption, industry-academia partnership is essential, and education must adapt. These findings underline the need for ethical awareness and teamwork while highlighting the transformational possibilities of AI and Blockchain.

10.2 Envisioning a Transformative and Inclusive Future of Communication Education

A whole new paradigm must be adopted in order to envision a future of communication education that is inclusive, revolutionary, and supported by AI and Blockchain. AI-driven personalization customizes learning routes, encouraging personal development. By guaranteeing the veracity of accomplishments, Blockchain enables an open and transparent educational system. Through overcoming geographic boundaries and providing equal access to learners from all backgrounds, this synergy democratizes education (Ayad et al,..2023)

Modules on ethical communication, artificial intelligence, and Blockchain inculcate values while preparing students for employment using technology. Teamwork abilities are improved through collaborative projects that use AI technologies to imitate real-world dynamics. As Blockchain validates credentials, inclusivity thrives as underrepresented voices may be heard (Ayad et al,.2023).A peaceful future cultivates flexible graduates with moral compass by integrating Blockchain's trust base and AI's analytical capabilities. This idea requires continual cooperation between academics, business, and decision-makers, ultimately converting.

10.3 Final Remarks on AI and Blockchain Applications in Industrial Robotics

In conclusion, the use of Blockchain technology and AI to industrial robots offers a future characterized by unheard-of creativity and effectiveness. Combining cognitive AI capabilities with Blockchain's safe, open architecture has the enormous potential to completely transform businesses by streamlining production processes, improving supply chain management, and guaranteeing data integrity (Alexandre et al..,2023) Industrial robots powered by AI have the potential to significantly increase production, accuracy, and safety, while Blockchain provides reliable data sharing and ecosystem collaboration.

However, there are obstacles on this transforming trip that need for careful navigating. Proactive attention is needed for ethical issues, labor reskilling, cybersecurity, and regulatory alignment. Industry, academics, legislators, and the general public must work together to fully realize the advantages of AI-driven industrial robots in a Blockchain -secured environment (Vasco et al,..2023).

REFERENCES

Agerskov, S. (2023). *Asger Balle Pedersen, Roman Beck, Ethical guidelines for Blockchain system*. ECIS.

Bal, Afacan, Clardy, & Cakir. (2023). Inclusive Future Making: Building a Culturally Responsive Behavioral Support System at an Urban Middle School with Local Stakeholders. Academic Press.

Liu, Y., & Ren, L. (2023). *Artificial Intelligence Techniques for Joint Sensing and Localization in Future Wireless Networks*. Hindawi.

Lopes & Alexandre. (2023). An Overview of Blockchain Integration with Robotics and Artificial Intelligence. Cornell University.

Shneiderman. (2023). Bridging the gap between ethics and practice: Guidelines for Reliable, safe and trustworthy human-centred AI system. ACM Journals.

Suresh Babu, C. V. (2022). *Artificial Intelligence and Expert Systems*. Anniyappa Publication.

Suresh Babu, C. V. & Das, S. (2023). Impact of Blockchain Technology on the Stock Market. In K. Mehta, R. Sharma, & P. Yu (Eds.), *Revolutionizing Financial Services and Markets Through FinTech and Blockchain* (pp. 44-59). IGI Global. doi:10.4018/978-1-6684-8624-5.ch004

Suresh Babu, C. V., & Padma, R. (2023). Technology Transformation Through Skilled Teachers in Teaching Accountancy. In R. González-Lezcano (Ed.), *Advancing STEM Education and Innovation in a Time of Distance Learning* (pp. 211–233). IGI Global. doi:10.4018/978-1-6684-5053-6.ch011

Suresh Babu, C. V., & Rohan, B. (2003). Evaluation and Quality Assurance for Rapid E-Learning and Development of Digital Learning Resources. In Implementing Rapid E-Learning Through Interactive Materials Development. IGI Global. doi:10.4018/978-1-6684-4940-0.ch008

Chapter 4

Elevating Performance for Enhancing AI-Powered Humanoid Robots Through Innovation

Krishnamohan Reddy Kunduru

https://orcid.org/0009-0009-8060-6216

Department of Engineering Design, Overhead Door Corporation, Lewisville, USA

Yagya Dutta Dwivedi

https://orcid.org/0000-0002-5793-9364

Department of Aeronautical Engineering, Institute of Aeronautical Engineering, Hyderabad, India

R. Aruna

Department of Electronics and Communication Engineering, AMC Engineering College, Bengaluru, India

G. R. Thippeswamy

Department of Computer Science and Engineering, Don Bosco Institute of Technology, Bengaluru, India

Subramanian Selvakumar

Department of Electrical and Computer Engineering, Bahir Dar Institute of Technology, Ethiopia

M. Sudhakar

Department of Mechanical Engineering, Sri Sai Ram Engineering College, Chennai, India

ABSTRACT

This chapter explores strategies for enhancing the performance of AI-powered humanoid robots through innovation. It begins with an overview of the current landscape of humanoid robots, highlighting their diverse applications across industries. The review examines existing performance metrics and identifies areas for improvement. Subsequent sections delve into specific avenues for innovation,

DOI: 10.4018/979-8-3693-2399-1.ch004

including advancements in cognitive capabilities, motor skills, emotional intelligence, and human-robot interaction. Leveraging machine learning techniques for continuous improvement is also explored. Ethical considerations, such as privacy concerns and bias mitigation, are addressed, along with challenges associated with societal impact. The chapter concludes with case studies showcasing successful implementations of performance-enhancing strategies and outlines potential future directions for research and development in the field.

INTRODUCTION

AI-powered humanoid robots are a significant advancement in robotics, enabling them to mimic human behavior and perform tasks like healthcare assistance and manufacturing productivity enhancement. However, their performance is not without challenges, and this introduction lays the groundwork for exploring innovative strategies to improve their performance. Humanoid robots are being deployed in various sectors, including healthcare, education, and entertainment. They enhance patient care, provide personalized learning experiences, and are increasingly integrated into retail environments, customer service roles, and households. They augment healthcare professionals' capabilities, enhance patient outcomes, and provide companionship, enhancing patient care and engagement (Rossos et al., 2023).

To maximize the impact of humanoid robots, they need to improve their performance across various dimensions. Traditional metrics like accuracy, speed, and efficiency may not fully capture the complexity of human-robot interaction and real-world scenarios. Therefore, there is a growing focus on developing more nuanced performance metrics that consider adaptability, versatility, and user experience. AI algorithms are improving the cognitive capabilities of humanoid robots, enabling them to learn from experience, recognize patterns, and make informed decisions in dynamic environments. Innovations in perception and sensing technologies also enhance their understanding and interpretation of their surroundings, facilitating intuitive interaction with humans and objects. Further research is needed to refine these capabilities (Mukherjee et al., 2022).

Humanoid robots' performance is enhanced by improving their motor skills and physical interaction capabilities. Sensorimotor integration, dexterity, and adaptive control mechanisms enable safe manipulation and navigation in complex environments. Advances in materials science, robotics, and biomechanics provide new opportunities to improve agility, strength, and flexibility, enabling more natural and fluid movements in humanoid robots. Researchers are exploring the use of emotional intelligence in humanoid robot development. Sentiment analysis, empathy algorithms, and affective computing techniques enable robots to recognize emotional cues, tailor responses, and establish meaningful connections with users. This approach can enhance user engagement, foster trust, and facilitate empathetic interactions in humanoid robots (Hlee et al., 2023).

Human-robot interaction is crucial for the success of humanoid robots, involving natural language processing, gesture recognition, and behavior adaptation. Advances in collaboration and shared autonomy enhance this partnership. AI-powered humanoid robots present a complex challenge, but by improving cognitive abilities, motor skills, emotional intelligence, and interaction, researchers can unlock new applications. However, technical, ethical, and societal issues remain. Collaboration and sustained innovation can unlock the full potential of AI-powered robots, enhancing productivity and quality of life (Nguyen et al., 2022).

AI-powered humanoid robots are a significant advancement in robotics, combining AI capabilities with human-like behavior. These robots are designed to interact with humans and their environments in a manner resembling human interaction, offering various applications across industries. They use advanced sensors, cameras, and other sensory devices to perceive and interpret the world, analyze visual and auditory cues, recognize objects and people, and navigate complex environments with high autonomy. AI-powered robots are equipped with advanced algorithms that enable them to learn from experience, make decisions, and adapt to changing circumstances. Machine learning techniques like deep learning and reinforcement learning help them acquire new skills and improve performance over time (Karami et al., 2023).

AI-powered humanoid robots are increasingly being used in various industries, including healthcare, education, entertainment, hospitality, and manufacturing. They assist in patient care, rehabilitation, therapy, and education by providing personalized learning experiences. In entertainment and hospitality, they entertain guests, provide information, and offer assistance with room service and concierge services. In manufacturing, they enhance productivity by performing tasks like assembly, inspection, and logistics in collaboration with human workers. AI-powered humanoid robots are increasingly being used in various industries, including healthcare, education, entertainment, hospitality, and manufacturing. They assist in patient care, rehabilitation, therapy, and education by providing personalized learning experiences (He & Zhang, 2023). In entertainment and hospitality, they entertain guests, provide information, and offer assistance with room service and concierge services. In manufacturing, they enhance productivity by performing tasks like assembly, inspection, and logistics in collaboration with human workers.

Humanoid robots with enhanced performance are more versatile and adaptable in various environments and tasks. They have advanced cognitive capabilities, enabling them to learn new skills, handle unforeseen situations effectively, and operate safely around humans and objects. This flexibility reduces accident and injury risks, while improved motor skills and cognitive capabilities enhance reliability and robustness.

Investing in performance enhancement of humanoid robots can give organizations a competitive edge, attracting more customers and driving innovation (Maheswari et al., 2023; Srinivas et al., 2023). This continuous improvement ensures they stay ahead of technological advancements and evolving customer demands (Khalid et al., 2024). As humanoid robots become more integrated into society, their long-term viability and sustainability are crucial. Performance enhancement not only improves current robot capabilities but alsolays the foundation for future advancements and innovation. This contributes to the widespread adoption of AI-powered humanoid robots in various industries (Liu et al., 2022).

Objectives

- Explore the current landscape of AI-powered humanoid robots, including their applications and capabilities.
- Identify key challenges and limitations in the performance of humanoid robots.
- Discuss the importance of performance enhancement in advancing the capabilities of humanoid robots.
- Present strategies and methodologies for enhancing the performance of AI-powered humanoid robots.
- Examine case studies and examples showcasing successful implementations of performance enhancement techniques.
- Highlight future directions and areas for further research and development in the field of humanoid robot performance enhancement.

Scope

i.   This chapter will provide an overview of the current state of AI-powered humanoid robots, discussing their applications across various industries and domains. It will explore the capabilities of humanoid robots, including their cognitive, motor, and interaction skills.

ii. The chapter will identify key challenges and limitations in the performance of AI-powered humanoid robots. This includes issues related to cognitive abilities, motor skills, human-robot interaction, and adaptability in dynamic environments.

iii. It will discuss why performance enhancement is crucial for maximizing the potential of humanoid robots. This includes improving efficiency, enhancing user experience, ensuring safety, and gaining a competitive advantage in the market.

iv. The chapter will present various strategies and methodologies for enhancing the performance of AI-powered humanoid robots. This may include advancements in AI algorithms, sensorimotor integration, emotional intelligence, and human-robot interaction techniques.

v. It will include case studies and examples showcasing successful implementations of performance enhancement techniques in real-world applications. This provides practical insights into the effectiveness of different approaches in improving humanoid robot performance.

vi. The chapter will discuss potential future directions and areas for further research and development in the field of humanoid robot performance enhancement. This includes emerging technologies, novel approaches, and challenges that need to be addressed to continue advancing the capabilities of humanoid robots.

The chapter provides a thorough understanding of performance enhancement in AI-powered humanoid robots, offering practical strategies for achieving this goal, based on its objectives and scope.

CURRENT LANDSCAPE AND PERFORMANCE METRICS

Humanoid robots have diverse applications across various industries, utilizing their unique capabilities to tackle specific needs and challenges. Humanoid robots are revolutionizing various aspects of society, including healthcare, education, customer service, and entertainment. As technology advances, new use cases emerge, transforming our lives and interactions with the world. As research and development progress, humanoid robots are expected to play a more prominent role in various aspects of society (Khalid et al., 2024; Martínez-Rojas et al., 2021).

- **Healthcare Assistance**: Humanoid robots are increasingly being utilized in healthcare settings to assist patients and healthcare professionals. They can provide companionship to elderly individuals, remind patients to take medication, and assist with activities of daily living such as getting out of bed or walking. In hospitals, humanoid robots can deliver supplies, assist with patient transport, and even perform basic medical procedures under supervision.

- **Education and Tutoring**: In educational settings, humanoid robots serve as interactive tutors and learning companions. They can engage students in personalized learning activities, provide feedback on assignments, and adapt their teaching strategies to individual learning styles. Humanoid robots are particularly beneficial for students with special needs or learning disabilities, providing them with additional support and encouragement (Prabhuswamy et al., 2024; D. M. Sharma et al., 2024; Venkatasubramanian et al., 2024).

- **Customer Service and Hospitality**: Humanoid robots are being deployed in hotels, airports, and other hospitality settings to enhance customer service and guest experience. They can greet guests, provide information about amenities and services, and assist with check-in and check-out procedures. In restaurants, humanoid robots may serve as waiters or waitresses, delivering food and drinks to tables and interacting with customers (Revathi et al., 2024).

- **Manufacturing and Logistics**: Humanoid robots play a crucial role in manufacturing and logistics operations, where they assist with tasks such as assembly, packaging, and material handling. They can operate alongside human workers on production lines, performing repetitive or physically demanding tasks with precision and efficiency. In warehouses, humanoid robots can autonomously navigate shelves, pick and pack orders, and transport goods to designated locations.

- **Entertainment and Media**: Humanoid robots are also employed in the entertainment industry for various purposes, including amusement park attractions, interactive exhibits, and live performances. They can entertain audiences with music, dance, and interactive games, providing immersive and memorable experiences. In media production, humanoid robots may be used as actors or presenters, delivering scripted lines or engaging in improvisational performances.

- **Research and Development**: Humanoid robots are valuable tools for research and development in fields such as robotics, artificial intelligence, and human-robot interaction. Researchers use humanoid robots to study human behavior, test new algorithms and control strategies, and explore novel applications and technologies. They serve as platforms for experimentation and innovation, pushing the boundaries of what is possible in robotics and AI.

Current Landscape of AI-Powered Humanoid Robots

AI-powered humanoid robots are rapidly advancing in various industries, including healthcare, education, manufacturing, retail, and hospitality. These robots, equipped with advanced algorithms, sensors, and actuators, perform tasks with human-like dexterity and intelligence. They enhance efficiency, improve user experiences, and enhance human capabilities. They play a crucial role in patient care, rehabilitation, therapy, education, automation, quality control, personalized learning, and companionship. They also enhance production efficiency and provide companionship to users in various contexts (He & Zhang, 2023).

Evaluation of Existing Performance Metrics

The current performance metrics for humanoid robots are primarily quantitative, focusing on accuracy, speed, efficiency, and reliability. However, these metrics may not fully capture the complexity of human-robot interaction and real-world scenarios. A growing recognition is for more nuanced metrics that consider adaptability, versatility, and user experience. Humanoid robot performance requires both quantitative and qualitative assessments, including user satisfaction, acceptance, and trust. Tools like human-robot interaction studies, usability testing, and user feedback can help evaluate subjective aspects and identify areas for improvement.

ADVANCEMENTS IN COGNITIVE CAPABILITIES

Humanoid robots can enhance their perception and sensing capabilities by gathering rich sensory information from their environment. This improves their interaction with humans and objects, facilitating seamless collaboration. AI-powered robots, with advanced algorithms and sensing technologies, can emulate human-like cognitive abilities, enabling greater autonomy and effectiveness in real-world environments, enhancing their performance (Chevalier et al., 2020).

Figure 1. Humanoid robots use sensory information

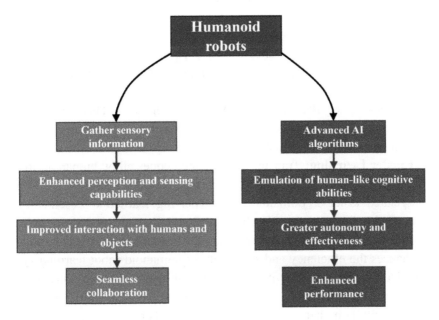

Figure 1 shows how humanoid robots use sensory information to improve their perception and sense, enhancing interaction with humans and objects. Advanced AI algorithms enable these robots to emulate human cognitive abilities, enhancing their autonomy and effectiveness in real-world environments, ultimately improving their performance.

Learning Algorithms

Learning algorithms, including deep learning, reinforcement learning, transfer learning, online learning, and hybrid approaches, are crucial in enhancing the cognitive capabilities of AI-powered humanoid robots. These algorithms enable robots to learn from experience, recognize patterns, make decisions, and adapt to new situations, mirroring human cognition. These advancements can be applied in various fields, such as healthcare, education, manufacturing, and service robotics, enhancing autonomy and efficiency in complex tasks (Karami et al., 2023; Khalid et al., 2024).

- **Deep Learning**: Deep learning techniques, such as artificial neural networks with multiple layers, have significantly improved the ability of humanoid robots to process and understand complex data. By training on large datasets, deep learning algorithms can extract high-level features and representations from raw sensor data, enabling robots to recognize objects, understand speech, and interpret visual scenes with greater accuracy and efficiency (Boopathi, 2023; Kav et al., 2023).
- **Reinforcement Learning**: Reinforcement learning is another powerful approach for training humanoid robots to perform complex tasks through trial and error. By rewarding desirable behaviors and penalizing undesirable ones, reinforcement learning algorithms enable robots to learn optimal strategies for achieving specified goals in dynamic environments. This flexibility makes reinforcement learning well-suited for tasks such as navigation, manipulation, and interaction with humans.
- **Transfer Learning**: Transfer learning techniques allow humanoid robots to leverage knowledge acquired from one task or domain to improve performance in related tasks or domains. By transferring learned representations and knowledge from one context to another, robots can adapt more quickly to new environments and tasks, reducing the need for extensive retraining. This enhances the efficiency and scalability of humanoid robot learning systems, particularly in scenarios with limited training data or computational resources.
- **Online Learning**: Online learning algorithms enable humanoid robots to incrementally update their knowledge and skills based on real-time data and

feedback. By continuously learning from interactions with the environment and users, robots can adapt to changing conditions and preferences, improving their performance over time. Online learning is particularly beneficial for applications requiring rapid adaptation, such as personalized assistance and collaborative tasks.

- **Hybrid Approaches**: Hybrid approaches that combine multiple learning algorithms, such as deep reinforcement learning or transfer learning with symbolic reasoning, offer synergistic benefits for enhancing the cognitive capabilities of humanoid robots. By integrating different learning paradigms, robots can leverage the strengths of each approach to address diverse challenges and achieve more robust and versatile performance across a wide range of tasks and domains.

Cognitive capabilities, including reasoning, decision-making, perception, and sensing, are crucial for AI-powered humanoid robots to improve their performance, enabling them to perceive, understand, and respond more effectively to their environment and tasks.

Reasoning and Decision-Making

The integration of advanced AI algorithms like machine learning, probabilistic reasoning, and symbolic reasoning is being used to enhance the reasoning and decision-making abilities of humanoid robots, enabling them to analyze complex situations and make informed choices (Hlee et al., 2023; Nguyen et al., 2022).

- **Machine Learning Algorithms:** Machine learning techniques, such as deep learning and reinforcement learning, enable humanoid robots to learn from experience and data, allowing them to recognize patterns, predict outcomes, and optimize decision-making processes. By training on large datasets, robots can acquire knowledge and expertise in specific domains, improving their ability to make accurate and timely decisions (Maheswari et al., 2023; Veeranjaneyulu et al., 2023).
- **Probabilistic Reasoning:** Probabilistic reasoning techniques, such as Bayesian inference and probabilistic graphical models, enable robots to reason under uncertainty, incorporating probabilistic information into decision-making processes. This allows robots to assess the likelihood of different outcomes and select actions that maximize expected utility or achieve specified objectives.
- **Symbolic Reasoning:** Symbolic reasoning techniques, such as logic-based reasoning and knowledge representation, enable robots to manipulate

symbols and logical rules to perform tasks that require abstract reasoning and problem-solving. This includes tasks such as planning, scheduling, and logical inference, where robots must generate and execute sequences of actions to achieve desired goals.

Humanoid robots are improving their reasoning and decision-making abilities, enabling them to navigate complex environments, adapt to changing situations, and perform tasks more efficiently.

Perception and Sensing

Improved perception and sensing technologies are crucial for humanoid robots to accurately interpret their environment, gather relevant information, and make informed decisions (Rossos et al., 2023).

- **Visual Perception:** Advancements in computer vision techniques enable humanoid robots to perceive and understand visual information from cameras and sensors. This includes tasks such as object recognition, scene understanding, depth estimation, and tracking, which are essential for navigating environments, interacting with objects, and recognizing human gestures and expressions.
- **Auditory Perception:** Auditory perception technologies enable robots to perceive and interpret sound signals from microphones and sensors. This includes tasks such as speech recognition, sound localization, and environmental sound classification, which are crucial for understanding spoken commands, detecting events, and interacting with humans in noisy environments.
- **Tactile Sensing:** Tactile sensing technologies enable robots to perceive and interpret tactile information from touch sensors and tactile arrays. This includes tasks such as object manipulation, grasping, and haptic feedback, which are essential for interacting with objects and humans in physical environments.

REFINING MOTOR SKILLS AND PHYSICAL INTERACTION

AI-powered humanoid robots need to improve their motor skills and physical interaction capabilities to perform tasks with precision, dexterity, and agility. This involves improving their ability to manipulate objects, navigate environments, and interact with humans and other robots safely, efficiently, and naturally. Advancements in sensorimotor integration enable robots to perceive and respond to their environment in real-time, coordinate their movements, and adapt to dynamic changes (Vemuri, 2023).

Figure 2. Process of improving motor skills and physical interaction capabilities

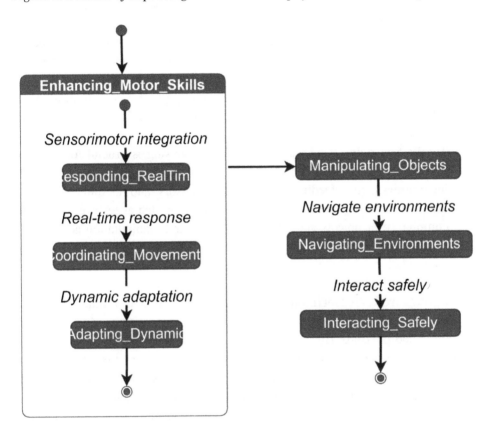

Figure 2 shows the process of improving motor skills and physical interaction capabilities in AI-powered humanoid robots. This includes enhancing object manipulation, navigation, and safe interaction. Sensorimotor integration, real-time response, and dynamic adaptation are also essential for efficient and natural interaction with the environment.

Sensorimotor Integration

Sensorimotor integration is the process of integrating sensory information from various sensors with motor commands to guide movements and actions. It is crucial in humanoid robots for enabling them to interact with their environment and perform tasks that require coordination between perception and action. This involves integrating sensors like cameras, depth sensors, force/torque sensors, and proprioceptive sensors (Moretti et al., 2020).

- **Visual-Sensorimotor Integration:** Visual information plays a central role in guiding the movements of humanoid robots. By integrating visual input from cameras and depth sensors with motor commands, robots can perceive objects, obstacles, and landmarks in their environment and plan appropriate actions to interact with them. Visual-sensorimotor integration enables tasks such as object manipulation, navigation, and gesture recognition, allowing robots to interact with humans and objects in a manner that is visually guided and contextually aware.
- **Tactile-Sensorimotor Integration:** Tactile sensing is crucial for enabling robots to interact with objects and humans in physical environments. By integrating tactile feedback from touch sensors and tactile arrays with motor commands, robots can perceive properties such as texture, shape, and compliance of objects and adjust their grasp and manipulation accordingly. Tactile-sensorimotor integration enables tasks such as object grasping, manipulation, and haptic exploration, facilitating more dexterous and precise interactions with the environment.
- **Force/Torque-Sensorimotor Integration:** Force and torque sensing is essential for enabling robots to interact with objects with varying degrees of force and pressure. By integrating force/torque feedback from sensors with motor commands, robots can adjust their movements and exert appropriate forces during manipulation tasks. Force/torque-sensorimotor integration enables tasks such as assembly, insertion, and manipulation of objects with varying degrees of compliance, facilitating more robust and adaptive interactions in industrial and collaborative settings (Boopathi, 2023; Boopathi & Kanike, 2023).

Sensorimotor integration refinements enable humanoid robots to perceive and respond to their environment with greater accuracy, adaptability, and efficiency. This improves their ability to perform complex tasks in diverse real-world environments. As sensorimotor integration advances, humanoid robots will exhibit more sophisticated motor skills and physical interaction capabilities, leading to new applications in healthcare, manufacturing, and service robotics.

Dexterity and Manipulation

Dexterity, the agility, precision, and coordination of humanoid robots, is crucial for tasks like assembly, manufacturing, household chores, and surgical procedures, and is being enhanced through advancements (Chevalier et al., 2020).

- **Gripping and Grasping Mechanisms:** Enhancements in gripping and grasping mechanisms enable robots to securely hold and manipulate objects of various shapes, sizes, and textures. This includes the development of versatile grippers, adaptive grasp planning algorithms, and tactile sensors that provide feedback on contact forces and object properties. By improving grip stability and adaptability, robots can handle fragile objects, grasp irregular shapes, and adjust their grip strength based on task requirements.
- **Fine Motor Control:** Advancements in fine motor control enable robots to perform delicate and precise movements with their manipulators and end-effectors. This includes improving joint flexibility, accuracy, and responsiveness, allowing robots to perform tasks that require intricate manipulation, such as threading needles, writing, or assembling small components. Fine-tuning control algorithms and feedback mechanisms enable robots to execute smooth and coordinated movements, mimicking human-like dexterity and precision.
- **Hand-Eye Coordination:** Enhancing hand-eye coordination enables robots to synchronize their manipulative actions with visual feedback from cameras and sensors. This involves developing algorithms for visual servoing, object tracking, and hand-eye calibration, enabling robots to accurately align their manipulators with target objects and perform tasks requiring precise spatial alignment. By integrating visual feedback into manipulation tasks, robots can compensate for errors, adjust their movements in real-time, and improve overall accuracy and efficiency.
- **Adaptive Manipulation Strategies:** Advancements in adaptive manipulation strategies enable robots to adjust their grasping and manipulation techniques based on the properties of the object and the task requirements. This includes developing algorithms for tactile sensing, object recognition, and grasp planning, allowing robots to select appropriate grasps and manipulation strategies for different objects and scenarios. By adapting their manipulation techniques to the task at hand, robots can handle novel objects, overcome uncertainties, and improve task performance in dynamic environments(Kumara et al., 2023).

AI-powered humanoid robots are enhancing their agility, precision, and effectiveness by refining motor skills and physical interaction capabilities. These advancements enable robots to manipulate objects, interact with humans, and navigate complex environments, paving the way for automation, assistance, and collaboration in various industries.

Adaptive Control

Adaptive control is a control theory that allows systems to adjust their parameters or behavior based on environmental feedback. It is particularly important in humanoid robots, as it enables them to respond dynamically to changes in their environment, ensuring robust and efficient performance (Martínez-Rojas et al., 2021).

- **Model-Free Adaptive Control:** Model-free adaptive control techniques, such as reinforcement learning and adaptive dynamic programming, enable robots to learn control policies directly from interaction with the environment, without explicitly modeling its dynamics. These techniques allow robots to adapt their control strategies based on trial-and-error learning, gradually improving performance over time through experience.
- **Model-Based Adaptive Control:** Model-based adaptive control techniques utilize mathematical models of the robot and its environment to predict system behavior and adjust control parameters accordingly. Adaptive model predictive control, for example, combines predictive models with online parameter adaptation to optimize control actions in real-time, ensuring robust performance in the face of uncertainties and disturbances.
- **Adaptive Compliance Control:** Adaptive compliance control techniques enable robots to adjust their stiffness and compliance parameters to interact safely and effectively with humans and objects. By modulating their stiffness in response to external forces and constraints, robots can achieve gentle and precise interaction with delicate objects or humans, while maintaining stability and safety.
- **Adaptive Trajectory Tracking:** Adaptive trajectory tracking techniques enable robots to track desired trajectories or motion profiles accurately, even in the presence of uncertainties or disturbances. By continuously updating trajectory parameters based on feedback from sensors, robots can compensate for deviations and disturbances, ensuring smooth and accurate motion execution in tasks such as grasping, manipulation, and locomotion.
- **Learning from Demonstration:** Learning from demonstration techniques enable robots to acquire new motor skills and interaction behaviors by observing and imitating human demonstrations. By leveraging techniques

such as imitation learning and apprenticeship learning, robots can learn complex manipulation tasks and interaction behaviors from human experts, adapting their control strategies to mimic human-like performance.

AI-powered humanoid robots utilize adaptive control techniques to enhance their dynamic navigation and interaction with their environment, ensuring robust and efficient performance in various tasks. These techniques allow robots to adapt their control strategies in real-time, enhancing seamless interaction with humans and objects.

INTEGRATION OF EMOTIONAL INTELLIGENCE

The integration of emotional intelligence in AI-powered humanoid robots involves recognizing and interpreting human emotions through sentiment analysis. This technique allows robots to interpret speech, gestures, and facial expressions, enabling them to tailor their responses and behaviors to better meet users' emotional needs, resulting in more empathetic and engaging interactions (Marcos-Pablos & García-Peñalvo, 2022).

Figure 3. The process of integrating emotional intelligence in AI-powered humanoid robots

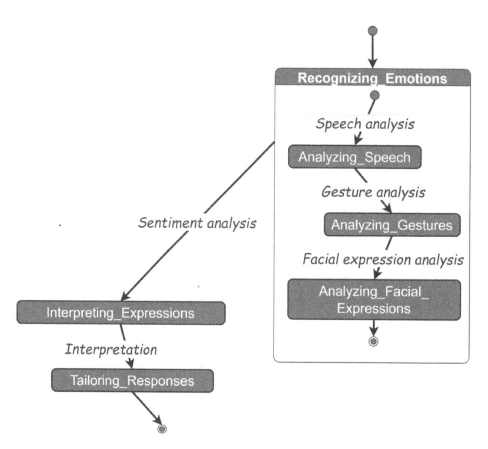

Figure 3 illustrates the process of integrating emotional intelligence in AI-powered humanoid robots. It begins with recognizing emotions through sentiment analysis, which involves analyzing speech, gestures, and facial expressions. The state of recognizing emotions further involves specific analyses such as speech analysis, gesture analysis, and facial expression analysis. Finally, the process leads to interpreting expressions and tailoring responses to meet users' emotional needs effectively, resulting in more empathetic and engaging interactions.

Sentiment Analysis

Sentiment analysis, also known as emotion recognition or affective computing, is a branch of artificial intelligence that analyzes and interprets human emotions through text, speech, and images. It helps AI-powered humanoid robots recognize and respond to users' emotional states, enhancing human-robot interaction and user experience (Pepito et al., 2020). The integration of sentiment analysis techniques into AI-powered humanoid robots can improve their ability to understand and respond empathetically to users' emotions. This fosters more engaging, personalized interactions, improving user satisfaction and acceptance of humanoid robots in various applications. Empathy algorithms, particularly through sentiment analysis, represent a significant advancement in AI-powered humanoid robot development, enhancing the quality of human-robot interaction and fostering more meaningful and empathetic relationships between robots and humans.

- **Speech-based Sentiment Analysis:** Speech-based sentiment analysis techniques enable robots to analyze the emotional content of human speech, including tone, intonation, and prosody. By using techniques such as acoustic feature extraction, speech recognition, and emotion classification, robots can identify emotional states such as happiness, sadness, anger, and surprise from spoken utterances. This allows robots to adapt their responses and behaviors to better match the emotional tone of the conversation, fostering more natural and empathetic interactions.
- **Facial Expression Recognition:** Facial expression recognition techniques enable robots to analyze the emotional content of human facial expressions, such as smiles, frowns, and raised eyebrows. By using computer vision algorithms and facial feature detection techniques, robots can detect subtle changes in facial expressions and infer the underlying emotional states of their users. This allows robots to respond appropriately to users' emotional cues, adjusting their behavior and expressions to convey empathy and understanding.
- **Gesture-based Sentiment Analysis:** Gesture-based sentiment analysis techniques enable robots to interpret the emotional content of human gestures and body language. By analyzing spatial and temporal features of gestures, robots can infer emotional states such as happiness, sadness, or frustration from hand movements, postures, and gestures. This allows robots to respond appropriately to users' emotional cues, adjusting their gestures and body language to convey empathy and support.
- **Multimodal Sentiment Analysis:** Multimodal sentiment analysis techniques integrate information from multiple modalities, such as speech, facial expressions, and gestures, to infer users' emotional states more accurately. By combining information from different sources, robots can improve the robustness and reliability of emotion recognition, enabling more nuanced and context-aware responses to users' emotional cues.

Empathy Algorithms

Empathy algorithms are AI-driven tools that enable robots to understand and respond to human emotions in a empathetic and socially appropriate manner, using techniques from affective computing, natural language processing, and computer vision (Pepito et al., 2020). Empathy algorithms are a significant advancement in AI-powered humanoid robots, allowing them to recognize, understand, and respond to human emotions. These algorithms, utilizing affective computing, natural language processing, and computer vision, improve human-robot interaction quality and foster more empathetic relationships between robots and humans. Affective computing focuses on recognizing, interpreting, and responding to human emotions effectively, enhancing the quality of human-robot interaction. This integration involves various techniques and methodologies to enable robots to perceive, understand, and respond to human emotions accurately and appropriately.

- **Emotion Recognition:** Empathy algorithms enable robots to recognize and interpret human emotions based on various modalities, including facial expressions, vocal intonations, body language, and physiological signals. Computer vision techniques, such as facial recognition and expression analysis, enable robots to detect subtle changes in facial expressions and infer underlying emotions. Similarly, audio processing algorithms analyze vocal cues, intonations, and speech patterns to identify emotional states, while sensors can detect physiological signals such as heart rate and skin conductivity to provide additional context.
- **Emotion Understanding:** Once emotional cues are detected, empathy algorithms enable robots to understand the underlying emotional states and their significance in the context of human-robot interaction. Natural language processing techniques analyze verbal communication to extract emotional content and sentiment, while context-aware algorithms consider situational factors and previous interactions to interpret emotional cues more accurately. By understanding the emotional context, robots can tailor their responses and behaviors to better meet the emotional needs of humans.
- **Emotion Expression:** Empathy algorithms also enable robots to express emotions in a manner that is natural and empathetic, enhancing the quality of human-robot interaction. Robots can use facial expressions, vocal intonations, gestures, and other non-verbal cues to convey empathy, compassion, and understanding. By mimicking human-like emotional expressions, robots can establish rapport with users and facilitate more engaging and effective communication.

- **Adaptive Behavior:** Empathy algorithms allow robots to adapt their behavior dynamically based on the emotional cues and feedback received from humans. For example, if a robot detects that a user is experiencing distress or frustration, it can adjust its behavior to provide comfort, reassurance, or assistance accordingly. By responding empathetically to human emotions, robots can enhance user satisfaction, trust, and engagement in various applications, such as healthcare, education, customer service, and companionship.

Affective Computing

Affective computing is a method that uses facial expressions, vocal intonation, physiological signals, and contextual information to understand and respond to human emotions. It helps robots adapt their behavior based on user's emotional state, enhancing their social intelligence and empathy, thereby fostering trust, rapport, and effective communication in humanoid robots (Abdollahi et al., 2022).

- Facial expression recognition techniques enable robots to analyze facial cues such as smiles, frowns, and eye movements to infer the emotional state of users. Machine learning algorithms, such as convolutional neural networks, are commonly used to detect and classify facial expressions accurately. By recognizing subtle changes in facial expressions, robots can gauge the emotional state of users and adjust their behavior accordingly to convey empathy and understanding.
- Speech emotion recognition techniques enable robots to analyze vocal intonation, speech rate, and other acoustic features to infer the emotional state of users from their speech. Machine learning algorithms, such as deep neural networks and support vector machines, are trained on labeled speech datasets to classify emotions accurately. By analyzing speech patterns, robots can detect emotions such as joy, sadness, anger, and surprise, enabling them to respond appropriately to users' emotional cues.
- Physiological signal analysis techniques enable robots to monitor users' physiological signals, such as heart rate, skin conductance, and electroencephalography (EEG), to infer their emotional state indirectly. By correlating changes in physiological signals with emotional arousal, robots can infer users' emotional responses to stimuli and adapt their behavior accordingly. Biofeedback techniques may also be employed to regulate users' emotional states actively, promoting relaxation and stress reduction.
- Contextual understanding techniques enable robots to interpret situational context and social cues to infer users' emotional states more accurately. By considering factors such as conversation context, social norms, and cultural

differences, robots can tailor their responses to match users' emotional expressions and social expectations. Natural language processing techniques, sentiment analysis algorithms, and context-aware reasoning mechanisms are often employed to enhance robots' contextual understanding and adaptability.

The integration of affective computing techniques into AI-powered humanoid robots can enhance their ability to understand and respond to human emotions, fostering trust and effective communication. This integration is crucial for ensuring positive user experiences and promoting the acceptance and adoption of these robots in society (Kumara et al., 2023; Tirlangi et al., 2024).

OPTIMIZING HUMAN-ROBOT INTERACTION

Human-robot interaction (HRI) is crucial for seamless communication between humans and AI-powered robots. Natural Language Processing (NLP) enhances HRI by allowing robots to understand and generate human language, facilitating intuitive communication. Integrating NLP techniques into robots allows them to interpret commands, answer questions, engage in dialogue, and provide assistance effectively, improving user experience and enabling robots to perform various tasks (Abdollahi et al., 2022).

Figure 4. Integration of natural language processing (NLP) techniques into AI-powered robots

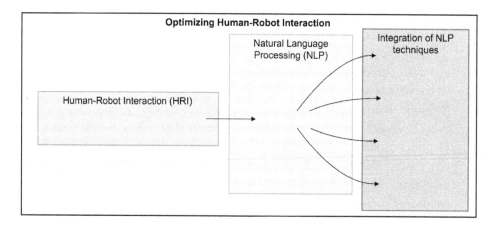

Figure 4 illustrates the integration of Natural Language Processing (NLP) techniques into AI-powered robots, focusing on optimizing human-robot interaction for seamless communication. NLP enhances HRI, allowing robots to understand and generate human language, enabling them to interpret commands, answer questions, engage in dialogue, and provide assistance, ultimately improving user experience and enabling robots to perform various tasks.

Natural Language Processing (NLP)

Natural Language Processing (NLP) is a method that uses computational linguistics, machine learning, and artificial intelligence to understand, interpret, and generate human language. It aids in tasks like speech recognition, sentiment analysis, and text generation. In the context of humanoid robots, NLP techniques are crucial for facilitating intuitive and efficient communication with humans using natural language (Gasteiger et al., 2023).

- **Speech Recognition:** Speech recognition techniques enable robots to convert spoken language into text, allowing them to understand and interpret user commands and queries. Advanced machine learning algorithms, such as deep learning-based recurrent neural networks, are trained on large speech datasets to accurately transcribe spoken language into text, even in noisy environments. By integrating speech recognition capabilities, humanoid robots can interact with users verbally, enabling hands-free operation and enhancing accessibility for users with disabilities.
- **Language Understanding:** Language understanding techniques enable robots to analyze and interpret the meaning of user utterances, enabling them to extract relevant information and take appropriate actions. Natural language understanding (NLU) algorithms leverage semantic parsing, syntactic analysis, and named entity recognition to infer the intent and context of user queries, enabling robots to provide relevant responses and assistance. By understanding the semantics and context of user input, humanoid robots can engage in more meaningful and contextually relevant interactions with users (M. Sharma et al., 2024; Sonia et al., 2024; Upadhyaya et al., 2024).
- **Dialog Management:** Dialog management techniques enable robots to engage in multi-turn conversations with users, maintaining context and coherence throughout the interaction. Dialog systems employ techniques such as state tracking, dialogue policy learning, and response generation to manage the flow of conversation, handle user requests, and generate appropriate responses. By maintaining context and continuity in dialogue,

humanoid robots can engage in more natural and fluid interactions with users, fostering engagement and facilitating effective communication.

- **Sentiment Analysis:** Sentiment analysis techniques enable robots to analyze and interpret the emotional tone and sentiment of user input, enabling them to respond appropriately to users' emotional cues. Machine learning algorithms, such as sentiment classifiers and emotion recognition models, analyze textual and contextual features to infer users' emotional states and adjust robot behavior accordingly. By recognizing and responding to users' emotional expressions, humanoid robots can provide empathetic and supportive interactions, enhancing user satisfaction and engagement.

- **Text Generation:** Text generation techniques enable robots to generate natural language responses and explanations, enabling them to provide informative and helpful responses to user queries. Natural language generation (NLG) algorithms leverage machine learning and language modeling techniques to generate human-like text based on input data and user preferences. By generating informative and contextually relevant responses, humanoid robots can enhance the quality of user interactions and provide valuable assistance in various domains.

The integration of Natural Language Processing (NLP) techniques into AI-powered humanoid robots can improve human-robot interaction by enabling them to understand and interpret user input. This enhances the overall user experience and allows for meaningful dialogue. As humanoid robots evolve, optimizing interaction through NLP techniques is crucial for their full potential. Gesture recognition technology is also essential in HRI, allowing robots to accurately interpret and respond to human gestures, facilitating intuitive and natural interaction in diverse settings.

Gesture Recognition

Gesture recognition technology uses sensors, cameras, and machine learning algorithms to interpret human gestures and body movements, accurately inferring users' intentions and preferences. This enhances human-robot interaction (HRI) by facilitating communication, controlling robot behavior, and enabling collaborative tasks, thus improving overall user experience (Aydin et al., 2020).

- **Gesture Detection:** Gesture recognition systems detect and identify human gestures using sensors such as cameras, depth sensors, and motion sensors. These sensors capture human movements and convert them into digital signals, which are processed by gesture recognition algorithms to identify specific gestures or patterns of movement. Machine learning techniques, such as convolutional neural networks and hidden Markov models, are commonly used to classify gestures accurately based on training data.

- **Gesture Classification:** Once gestures are detected, gesture recognition algorithms classify them into predefined categories or commands based on their characteristics and context. Common gestures include hand movements, arm gestures, facial expressions, and body postures, each of which may convey different meanings or commands to the robot. By training gesture recognition models on labeled datasets, robots can learn to recognize and classify a wide range of gestures accurately, enabling more intuitive and versatile interaction with users.
- **Gesture Mapping:** Gesture recognition systems map recognized gestures to corresponding robot actions or responses, enabling robots to interpret users' intentions and execute appropriate behaviors. This involves defining mappings between detected gestures and predefined robot commands or behaviors, such as moving forward, turning, grasping objects, or expressing emotions. By mapping gestures to specific actions, robots can respond dynamically to users' gestures, enabling seamless and intuitive HRI experiences.
- **Gesture Feedback:** Gesture recognition systems provide feedback to users to confirm that their gestures have been recognized and understood by the robot. This may involve visual feedback, such as displaying a gesture recognition icon or animation on a screen or robot display, auditory feedback, such as playing a sound or speech response, or haptic feedback, such as vibrating or moving the robot's body in response to recognized gestures. Providing feedback enhances the transparency and usability of gesture-based interaction, enabling users to communicate effectively with the robot.

Glide recognition technology in AI-powered humanoid robots can enhance human-robot interaction (HRI) experiences by accurately interpreting user intentions and preferences. This technology is used in healthcare, education, retail, and entertainment. As technology advances, robots will become more proficient in interpreting and responding to human gestures, enhancing their capabilities and society's acceptance. HRI optimization involves communication, gesture recognition, and behavior adaptation.

Behavior Adaptation

AI-powered humanoid robots can dynamically adjust their behavior to suit changes in the environment, user preferences, and task requirements using behavior adaptation techniques, which utilize machine learning algorithms, decision-making frameworks, and feedback mechanisms (Schaefer et al., 2021).

a) **Machine Learning-Based Adaptation:** Machine learning algorithms, such as reinforcement learning and imitation learning, enable robots to adapt their behavior based on experience and feedback from the environment. Reinforcement learning techniques allow robots to learn optimal behavior policies through trial-and-error interaction with the environment, maximizing cumulative rewards over time. Imitation learning techniques enable robots to mimic human behavior by observing and imitating expert demonstrations, facilitating more natural and intuitive interaction with users (Boopathi, 2024; Maheswari et al., 2023).

b) **Contextual Adaptation:** Contextual adaptation techniques enable robots to adjust their behavior based on situational context and environmental cues. By considering factors such as location, time, social norms, and task requirements, robots can tailor their responses to match the specific needs and preferences of users. Context-aware reasoning mechanisms and decision-making frameworks are often employed to facilitate contextual adaptation and optimize human-robot interaction in diverse settings.

c) **Feedback-Based Adaptation:** Feedback-based adaptation techniques enable robots to adjust their behavior based on feedback from users and sensors. By soliciting feedback from users through explicit commands, gestures, or facial expressions, robots can learn from user preferences and adapt their behavior accordingly. Additionally, sensors such as cameras, microphones, and touch sensors provide real-time feedback on user engagement and satisfaction, enabling robots to adjust their behavior dynamically to optimize user experiences.

d) **Personalization and Customization:** Personalization and customization techniques enable robots to adapt their behavior to individual user preferences and characteristics. By maintaining user profiles and preferences, robots can personalize their interactions with users, providing tailored responses and recommendations based on past interactions and feedback. Customization mechanisms allow users to configure robot behavior according to their preferences, enabling a more personalized and satisfying user experience.

e) **Adaptive Dialogue Management:** Adaptive dialogue management techniques enable robots to adapt their conversational strategies based on user input, feedback, and conversational context. By dynamically adjusting speech rate, tone, and content, robots can maintain engaging and effective communication with users, facilitating natural and fluid conversation flow. Natural language processing algorithms and dialogue management frameworks are often employed to enable adaptive dialogue management in humanoid robots.

AI-powered humanoid robots can enhance user experiences by incorporating behavior adaptation techniques. These techniques allow robots to adjust their behavior dynamically based on user preferences, contextual cues, and feedback. This fosters natural, intuitive communication and collaboration between humans and robots. As humanoid robots become more integrated into daily life, behavior adaptation is crucial for promoting positive user experiences and widespread adoption.

LEVERAGING MACHINE LEARNING FOR CONTINUOUS IMPROVEMENT

Machine learning is crucial for continuously improving the performance of AI-powered humanoid robots. It enables robots to learn from data, adapt their behavior, and improve performance through iterative feedback loops. This process involves data-driven insights, adaptive learning algorithms, and dynamic performance monitoring to identify areas for improvement (Bharadiya, 2023). Figure 5 demonstrates how machine learning is used to enhance the performance of AI-powered humanoid robots. This process involves learning from data, adapting behavior, and improving performance through iterative feedback loops, data-driven insights, adaptive learning algorithms, and dynamic performance monitoring.

Figure 5. Machine learning for enhancing the performance of AI-powered humanoid robots

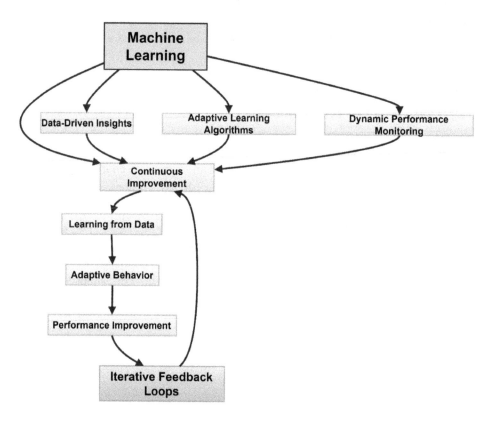

Data-Driven Insights

AI-powered humanoid robots utilize data-driven insights to improve their performance, user preferences, and task requirements. Machine learning algorithms like clustering, classification, and regression extract meaningful patterns from data, enabling robots to identify trends, anomalies, and opportunities for improvement, thereby enhancing their overall performance (Vaka et al., 2020).

- **User Interaction Data:** Data collected from user interactions, such as speech commands, gestures, and facial expressions, provide valuable insights into user preferences, engagement levels, and satisfaction. By analyzing user interaction data, robots can identify common patterns and adapt their behavior to better meet user needs and expectations.

- **Sensor Data:** Sensor data, including visual, auditory, and tactile feedback, provide valuable information about the robot's environment, task performance, and interaction dynamics. By analyzing sensor data, robots can detect environmental changes, identify obstacles or hazards, and optimize navigation, manipulation, and interaction strategies accordingly.
- **Performance Metrics:** Robot performance is evaluated through performance metrics like accuracy, speed, efficiency, and user satisfaction, which help assess progress, identify areas for improvement, and prioritize optimization efforts to achieve desired performance objectives.

Adaptive Learning

Adaptive learning techniques, including reinforcement learning, online learning, and transfer learning, enable robots to dynamically adjust their behavior based on environmental feedback and user interactions, enabling them to continuously update models and optimize performance in real-time (Shaikh et al., 2022).

- **Reinforcement Learning:** Reinforcement learning enables robots to learn optimal behavior policies through trial-and-error interaction with the environment, maximizing cumulative rewards over time. By receiving feedback in the form of rewards or penalties, robots can update their policy parameters and adjust their behavior to achieve desired objectives.
- **Online Learning:** Online learning techniques enable robots to learn from streaming data in real-time, continuously updating their models and adapting their strategies to changing circumstances. By processing data incrementally and updating model parameters iteratively, robots can adapt to dynamic environments and evolving user preferences efficiently.
- **Transfer Learning:** Transfer learning techniques enable robots to transfer knowledge and skills learned in one task or domain to related tasks or domains. By leveraging pre-trained models and transferring learned representations, robots can accelerate learning in new tasks, minimize data requirements, and adapt more quickly to novel scenarios.

Performance Monitoring

Performance monitoring mechanisms enable robots to track key metrics, evaluate against objectives, identify deviations, diagnose issues, and initiate corrective actions for iterative performance improvement (Vaka et al., 2020).

- **Real-Time Feedback:** Real-time feedback mechanisms provide immediate feedback on robot performance, enabling robots to adjust their behavior dynamically based on current performance levels and user feedback. By soliciting feedback from users or sensors in real-time, robots can identify performance issues promptly and take corrective actions to address them.
- **Performance Evaluation:** Performance evaluation techniques enable robots to assess their performance objectively against predefined performance metrics and benchmarks. By comparing actual performance metrics with desired performance levels, robots can identify areas for improvement and prioritize optimization efforts effectively (Gunasekaran et al., 2022; Vignesh et al., 2018).
- **Continuous Improvement:** Continuous improvement processes enable robots to iteratively optimize performance over time, leveraging feedback from user interactions, environmental feedback, and performance monitoring mechanisms. By adopting a cycle of data collection, analysis, adaptation, and evaluation, robots can continuously refine their models and strategies to achieve increasingly higher levels of performance and effectiveness.

AI-powered humanoid robots can adapt and evolve over time through machine learning, enhancing their capabilities and performance in diverse tasks. Data-driven insights, adaptive learning algorithms, and performance metrics optimize robot behavior, fostering continuous improvement and innovation in human-robot interaction.

CHALLENGES AND ETHICAL CONSIDERATIONS

The development and deployment of AI-powered humanoid robots require careful consideration of privacy, bias mitigation, and societal impact to ensure ethical and responsible use (Chevalier et al., 2020; Gasteiger et al., 2023; He & Zhang, 2023).

Privacy Concerns

Privacy concerns arise due to the collection, storage, and utilization of personal data by AI-powered humanoid robots. These robots often interact closely with users, collecting sensitive information such as speech, images, and behavioral patterns. To mitigate privacy risks, developers must implement robust privacy protection measures, such as data anonymization, encryption, and user consent mechanisms. Additionally, transparent data handling policies and clear communication with users regarding data usage are essential to building trust and ensuring user privacy.

Bias Mitigation

Bias in AI-powered humanoid robots can lead to unfair or discriminatory outcomes, particularly in decision-making processes. Biases may arise from biased training data, algorithmic design choices, or societal prejudices embedded in the development process. To mitigate bias, developers must carefully curate training data to ensure diversity and fairness, implement bias detection and mitigation techniques in algorithmic design, and conduct rigorous testing and validation to identify and address biases. Moreover, fostering diversity and inclusivity in the development team can help mitigate unconscious biases and ensure more equitable outcomes.

Societal Impact

The widespread use of AI-powered humanoid robots has significant societal implications, including economic, cultural, and ethical concerns. These robots may disrupt labor markets, challenge social norms, and raise questions about identity, autonomy, and human-robot relationships. To address these concerns, stakeholders must engage in inclusive dialogue and policymaking processes, ensure equitable distribution of benefits, and develop ethical frameworks for responsible design, deployment, and use. By addressing privacy concerns, mitigating biases, and considering societal impacts, stakeholders can foster trust, fairness, and inclusivity in human-robot interactions, maximizing the positive impact of humanoid robots while minimizing potential risks.

FUTURE DIRECTIONS

AI-powered humanoid robots have the potential to revolutionize industries and improve human well-being, but successful implementation and research require collaboration between researchers, developers, policymakers, and end-users (Khalid et al., 2024; Marcos-Pablos & García-Peñalvo, 2022).

Successful Implementations: Future directions for AI-powered humanoid robots will involve scaling successful implementations across various industries and domains. Identifying and replicating use cases where humanoid robots have demonstrated significant value, such as healthcare assistance, education, and customer service, will be crucial for widespread adoption. Furthermore, ongoing refinement and optimization of existing deployments will be necessary to enhance performance, reliability, and user acceptance.

Future Research Areas: The future of AI-powered humanoid robots will be influenced by ongoing research and development in several key areas.

- **Advanced AI Algorithms:** Further advancements in AI algorithms, including deep learning, reinforcement learning, and natural language processing, will enable robots to acquire more sophisticated cognitive abilities and adapt to a wider range of tasks and environments (Puranik et al., 2024).
- **Human-Robot Collaboration:** Research into human-robot collaboration will focus on enhancing the synergy between humans and robots in shared workspaces. This includes developing intuitive interfaces, collaborative task planning algorithms, and safety mechanisms to facilitate seamless cooperation and coordination (Puranik et al., 2024).
- **Ethical and Regulatory Frameworks:** The development of ethical and regulatory frameworks will be essential to govern the responsible design, deployment, and use of humanoid robots. This includes addressing concerns related to privacy, bias, safety, and accountability, and ensuring that robots adhere to ethical principles and societal norms.
- **Emotional Intelligence:** Advancements in affective computing and emotional intelligence will enable robots to better understand and respond to human emotions, enhancing the quality of human-robot interaction and fostering more empathetic and socially intelligent relationships.
- **Autonomous Learning and Adaptation:** Research into autonomous learning and adaptation will focus on enabling robots to acquire new skills and knowledge autonomously through interaction with their environment and feedback from users. This includes developing lifelong learning algorithms, self-supervised learning techniques, and adaptive control mechanisms.
- **Hardware Advancements:** Continued advancements in robotics hardware, including sensors, actuators, and materials, will enable the development of more agile, dexterous, and versatile humanoid robots. This includes improvements in energy efficiency, robustness, and miniaturization to enable robots to operate in diverse environments and perform complex tasks.
- **Human-Centered Design:** Human-centered design principles will guide the development of humanoid robots that prioritize user needs, preferences, and comfort. This includes designing robots with intuitive interfaces, naturalistic movement patterns, and customizable behaviors to enhance user acceptance and satisfaction.

CONCLUSION

AI-powered humanoid robots are a revolutionary technology that can revolutionize various aspects of human life and society. Equipped with advanced AI, sensors, and actuators, these robots possess human-like capabilities, allowing them to interact seamlessly with humans and their environment. This chapter explores the current landscape of AI-powered humanoid robots, discusses strategies for improving performance, and addresses ethical considerations in their development and deployment. The advancement of humanoid robots holds great promise for enhancing productivity, efficiency, and quality of life across various industries, but must be accompanied by careful consideration of privacy concerns, bias mitigation, and societal impact.

Humanoid robots' successful implementations in sectors like healthcare, education, and manufacturing offer valuable insights for future developments. Research in advanced AI algorithms, human-robot collaboration, and ethical frameworks will shape their evolution. Collaboration between researchers, developers, policymakers, and end-users is essential to realize the full potential of AI-powered robots. Addressing technical, ethical, and societal considerations can harness the transformative power of humanoid robots, creating a harmonious future where humans and robots coexist, enhancing human well-being and advancing technology.

ABBREVIATIONS

- **AI**: Artificial Intelligence
- **EEG**: Electroencephalography
- **HRI**: Human-Robot Interaction
- **NLP**: Natural Language Processing
- **NLU**: Natural Language Understanding
- **NLG**: Natural Language Generation

REFERENCES

Abdollahi, H., Mahoor, M., Zandie, R., Sewierski, J., & Qualls, S. (2022). Artificial emotional intelligence in socially assistive robots for older adults: A pilot study. *IEEE Transactions on Affective Computing*. PMID:37840968

Aydin, Y., Tokatli, O., Patoglu, V., & Basdogan, C. (2020). A computational multicriteria optimization approach to controller design for physical human-robot interaction. *IEEE Transactions on Robotics*, *36*(6), 1791–1804. doi:10.1109/TRO.2020.2998606

Bharadiya, J. P. (2023). Machine learning and AI in business intelligence: Trends and opportunities. [IJC]. *International Journal of Computer*, *48*(1), 123–134.

Boopathi, S. (2023). Deep Learning Techniques Applied for Automatic Sentence Generation. In Promoting Diversity, Equity, and Inclusion in Language Learning Environments (pp. 255–273). IGI Global. doi:10.4018/978-1-6684-3632-5.ch016

Boopathi, S. (2024). Advancements in Machine Learning and AI for Intelligent Systems in Drone Applications for Smart City Developments. In *Futuristic e-Governance Security With Deep Learning Applications* (pp. 15–45). IGI Global. doi:10.4018/978-1-6684-9596-4.ch002

Boopathi, S., & Kanike, U. K. (2023). Applications of Artificial Intelligent and Machine Learning Techniques in Image Processing. In *Handbook of Research on Thrust Technologies' Effect on Image Processing* (pp. 151–173). IGI Global. doi:10.4018/978-1-6684-8618-4.ch010

Chevalier, P., Kompatsiari, K., Ciardo, F., & Wykowska, A. (2020). Examining joint attention with the use of humanoid robots-A new approach to study fundamental mechanisms of social cognition. *Psychonomic Bulletin & Review*, 27(2), 217–236. doi:10.3758/s13423-019-01689-4 PMID:31848909

Gasteiger, N., Hellou, M., & Ahn, H. S. (2023). Factors for personalization and localization to optimize human–robot interaction: A literature review. *International Journal of Social Robotics*, 15(4), 689–701. doi:10.1007/s12369-021-00811-8

Gunasekaran, K., Boopathi, S., & Sureshkumar, M. (2022). Analysis of a Cryogenically Cooled Near-Dry WEDM Process using Different Dielectrics. *Mater-Tehnol.Si. Materials Technology*, 56(2), 179–186.

He, A.-Z., & Zhang, Y. (2023). AI-powered touch points in the customer journey: A systematic literature review and research agenda. *Journal of Research in Interactive Marketing*, 17(4), 620–639. doi:10.1108/JRIM-03-2022-0082

Hlee, S., Park, J., Park, H., Koo, C., & Chang, Y. (2023). Understanding customer's meaningful engagement with AI-powered service robots. *Information Technology & People*, 36(3), 1020–1047. doi:10.1108/ITP-10-2020-0740

Karami, V., Yaffe, M. J., & Rahimi, S. A. (2023). Early Detection of Alzheimer's Disease Assisted by AI-Powered Human-Robot Communication. In *Machine Learning and Artificial Intelligence in Healthcare Systems* (pp. 331–348). CRC Press.

KAV, R. P., Pandraju, T. K. S., Boopathi, S., Saravanan, P., Rathan, S. K., & Sathish, T. (2023). Hybrid Deep Learning Technique for Optimal Wind Mill Speed Estimation. *2023 7th International Conference on Electronics, Communication and Aerospace Technology (ICECA)*, 181–186.

Khalid, U., Naeem, M., Stasolla, F., Syed, M., Abbas, M., & Coronato, A. (2024). Impact of AI-Powered Solutions in Rehabilitation Process: Recent Improvements and Future Trends. *International Journal of General Medicine, 17,* 943–969. doi:10.2147/IJGM.S453903 PMID:38495919

Kumara, V., Sharma, M. D., Samson Isaac, J., Saravanan, S., Suganthi, D., & Boopathi, S. (2023). An AI-Integrated Green Power Monitoring System: Empowering Small and Medium Enterprises. In Advances in Environmental Engineering and Green Technologies (pp. 218–244). IGI Global. doi:10.4018/979-8-3693-0338-2.ch013

Liu, X., He, X., Wang, M., & Shen, H. (2022). What influences patients' continuance intention to use AI-powered service robots at hospitals? The role of individual characteristics. *Technology in Society, 70,* 101996. doi:10.1016/j.techsoc.2022.101996

Maheswari, B. U., Imambi, S. S., Hasan, D., Meenakshi, S., Pratheep, V., & Boopathi, S. (2023). Internet of things and machine learning-integrated smart robotics. In Global Perspectives on Robotics and Autonomous Systems: Development and Applications (pp. 240–258). IGI Global. doi:10.4018/978-1-6684-7791-5.ch010

Marcos-Pablos, S., & García-Peñalvo, F. J. (2022). Emotional intelligence in robotics: A scoping review. *New Trends in Disruptive Technologies. Tech Ethics and Artificial Intelligence: The DITTET Collection, 1,* 66–75.

Martínez-Rojas, A., Sánchez-Oliva, J., López-Carnicer, J. M., & Jiménez-Ramírez, A. (2021). Airpa: An architecture to support the execution and maintenance of AI-powered RPA robots. *International Conference on Business Process Management,* 38–48. 10.1007/978-3-030-85867-4_4

Moretti, C. B., Delbem, A. C., & Krebs, H. I. (2020). Human-robot interaction: Kinematic and kinetic data analysis framework. *2020 8th IEEE RAS/EMBS International Conference for Biomedical Robotics and Biomechatronics (BioRob),* 235–239.

Mukherjee, D., Gupta, K., & Najjaran, H. (2022). An ai-powered hierarchical communication framework for robust human-robot collaboration in industrial settings. *2022 31st IEEE International Conference on Robot and Human Interactive Communication (RO-MAN),* 1321–1326.

Nguyen, T.-H., Tran, D.-N., Vo, D.-L., Mai, V.-H., & Dao, X.-Q. (2022). AI-powered university: Design and deployment of robot assistant for smart universities. *Journal of Advances in Information Technology, 13*(1).

Pepito, J. A., Ito, H., Betriana, F., Tanioka, T., & Locsin, R. C. (2020). Intelligent humanoid robots expressing artificial humanlike empathy in nursing situations. *Nursing Philosophy*, *21*(4), e12318. doi:10.1111/nup.12318 PMID:33462939

Prabhuswamy, M., Tripathi, R., Vijayakumar, M., Thulasimani, T., Sundharesalingam, P., & Sampath, B. (2024). A Study on the Complex Nature of Higher Education Leadership: An Innovative Approach. In *Challenges of Globalization and Inclusivity in Academic Research* (pp. 202–223). IGI Global. doi:10.4018/979-8-3693-1371-8. ch013

Puranik, T. A., Shaik, N., Vankudoth, R., Kolhe, M. R., Yadav, N., & Boopathi, S. (2024). Study on Harmonizing Human-Robot (Drone) Collaboration: Navigating Seamless Interactions in Collaborative Environments. In Cybersecurity Issues and Challenges in the Drone Industry (pp. 1–26). IGI Global.

Revathi, S., Babu, M., Rajkumar, N., Meti, V. K. V., Kandavalli, S. R., & Boopathi, S. (2024). Unleashing the Future Potential of 4D Printing: Exploring Applications in Wearable Technology, Robotics, Energy, Transportation, and Fashion. In Human-Centered Approaches in Industry 5.0: Human-Machine Interaction, Virtual Reality Training, and Customer Sentiment Analysis (pp. 131–153). IGI Global.

Rossos, D., Mihailidis, A., & Laschowski, B. (2023). AI-powered smart glasses for sensing and recognition of human-robot walking environments. bioRxiv, 2023–10. doi:10.1101/2023.10.24.563804

Schaefer, S., Leung, K., Ivanovic, B., & Pavone, M. (2021). Leveraging neural network gradients within trajectory optimization for proactive human-robot interactions. *2021 IEEE International Conference on Robotics and Automation (ICRA)*, 9673–9679. 10.1109/ICRA48506.2021.9561443

Shaikh, T. A., Rasool, T., & Lone, F. R. (2022). Towards leveraging the role of machine learning and artificial intelligence in precision agriculture and smart farming. *Computers and Electronics in Agriculture*, *198*, 107119. doi:10.1016/j.compag.2022.107119

Sharma, D. M., Ramana, K. V., Jothilakshmi, R., Verma, R., Maheswari, B. U., & Boopathi, S. (2024). Integrating Generative AI Into K-12 Curriculums and Pedagogies in India: Opportunities and Challenges. *Facilitating Global Collaboration and Knowledge Sharing in Higher Education With Generative AI*, 133–161.

Sharma, M., Sharma, M., Sharma, N., & Boopathi, S. (2024). Building Sustainable Smart Cities Through Cloud and Intelligent Parking System. In *Handbook of Research on AI and ML for Intelligent Machines and Systems* (pp. 195–222). IGI Global.

Sonia, R., Gupta, N., Manikandan, K., Hemalatha, R., Kumar, M. J., & Boopathi, S. (2024). Strengthening Security, Privacy, and Trust in Artificial Intelligence Drones for Smart Cities. In *Analyzing and Mitigating Security Risks in Cloud Computing* (pp. 214–242). IGI Global. doi:10.4018/979-8-3693-3249-8.ch011

Srinivas, B., Maguluri, L. P., Naidu, K. V., Reddy, L. C. S., Deivakani, M., & Boopathi, S. (2023). Architecture and Framework for Interfacing Cloud-Enabled Robots. In *Handbook of Research on Data Science and Cybersecurity Innovations in Industry 4.0 Technologies* (pp. 542–560). IGI Global. doi:10.4018/978-1-6684-8145-5.ch027

Tirlangi, S., Teotia, S., Padmapriya, G., Senthil Kumar, S., Dhotre, S., & Boopathi, S. (2024). Cloud Computing and Machine Learning in the Green Power Sector: Data Management and Analysis for Sustainable Energy. In Developments Towards Next Generation Intelligent Systems for Sustainable Development (pp. 148–179). IGI Global. doi:10.4018/979-8-3693-5643-2.ch006

Upadhyaya, A. N., Saqib, A., Devi, J. V., Rallapalli, S., Sudha, S., & Boopathi, S. (2024). Implementation of the Internet of Things (IoT) in Remote Healthcare. In Advances in Medical Technologies and Clinical Practice (pp. 104–124). IGI Global. doi:10.4018/979-8-3693-1934-5.ch006

Vaka, A. R., Soni, B., & Reddy, S. (2020). Breast cancer detection by leveraging Machine Learning. *Ict Express*, *6*(4), 320–324. doi:10.1016/j.icte.2020.04.009

Veeranjaneyulu, R., Boopathi, S., Kumari, R. K., Vidyarthi, A., Isaac, J. S., & Jaiganesh, V. (2023). Air Quality Improvement and Optimisation Using Machine Learning Technique. *IEEE- Explore*, 1–6.

Vemuri, N. V. N. (2023). Enhancing Human-Robot Collaboration in Industry 4.0 with AI-driven HRI. *Power System Technology*, *47*(4), 341–358. doi:10.52783/pst.196

Venkatasubramanian, V., Chitra, M., Sudha, R., Singh, V. P., Jefferson, K., & Boopathi, S. (2024). Examining the Impacts of Course Outcome Analysis in Indian Higher Education: Enhancing Educational Quality. In Challenges of Globalization and Inclusivity in Academic Research (pp. 124–145). IGI Global.

Vignesh, S., Arulshri, K., SyedSajith, S., Kathiresan, S., Boopathi, S., & Dinesh Babu, P. (2018). Design and development of ornithopter and experimental analysis of flapping rate under various operating conditions. *Materials Today: Proceedings*, *5*(11), 25185–25194. doi:10.1016/j.matpr.2018.10.320

Chapter 5
Revolutionizing Friction Stir Welding With AI-Integrated Humanoid Robots

B. Shamreen Ahamed
Department of Computer Science and Engineering, Sathyabama Institute of Science and Technology, Chennai, India

R. Malkiya Rasalin Prince
Department of Mechanical Engineering, Karunya Institute of Technology and Sciences, Coimbatore, India

Katragadda Sudhir Chakravarthy
Department of Mechanical Engineering, PACE Institute of Technology and Sciences, India

S. Boopathi
Mechanical Engineering, Muthayammal Engineering College, Namakkal, India

Jeswin Arputhabalan
Department of Mechanical Engineering, Sri Sai Ram Institute of Technology, Chennai, India

S. Muthuvel
Department of Mechanical Engineering, Kalasalingam Academy of Research and Education, Srivilliputhur, India

K. Sasirekha
Department of Computer Science and Business Systems, R.M.D. Engineering College, Kavaraipattei, India

ABSTRACT

This chapter explores the use of AI-integrated humanoid robots in friction stir welding (FSW), a crucial process for joining materials without melting. By combining AI capabilities with humanoid robots' dexterity and adaptability, significant advancements can be achieved. AI algorithms can improve precision and accuracy by continuously analyzing real-time sensor data, while AI-powered

DOI: 10.4018/979-8-3693-2399-1.ch005

predictive maintenance can minimize downtime and enhance efficiency. AI-enabled robots in FSW increase automation, reduce human operator reliance, and minimize safety risks in hazardous environments. However, challenges such as cybersecurity concerns, regulatory hurdles, and ethical implications require careful consideration. Future research should focus on developing advanced AI algorithms, optimizing robot-human collaboration, and exploring new applications beyond traditional materials. The approach offers precision, efficiency, and safety, but necessitates interdisciplinary collaboration, strategic investment, and proactive addressing of technological, ethical, and regulatory challenges.

INTRODUCTION

The integration of artificial intelligence and robotics is revolutionizing traditional manufacturing processes, particularly in friction stir welding (FSW). This solid-state joining technique, used in various industries, is being enhanced by AI and robotics, leading to improved precision, efficiency, and safety, paving the way for unprecedented innovation in welding technology. Friction stir welding (FSW) is a highly effective method for producing high-quality welds with minimal defects in various materials like aluminum, steel, composites, and exotic alloys. Unlike traditional methods that involve melting base materials, FSW uses a non-consumable tool to mechanically stir and mix materials at the joint interface, resulting in superior mechanical properties and reduced distortion (Hunde & Woldeyohannes, 2022).

The introduction of humanoid robots with AI capabilities is a significant advancement in welding technology. These robots, designed to mimic human movements, provide flexibility and adaptability in complex tasks like FSW. By integrating AI algorithms, they achieve high precision and efficiency in welding processes. AI-integrated humanoid robots in FSW analyze and adapt to real-time data from sensors and feedback mechanisms, optimizing welding parameters like rotational speed, traverse rate, and applied force. They also enable predictive maintenance capabilities, detecting potential issues before they escalate, minimizing downtime and optimizing productivity, thus ensuring consistent weld quality across diverse materials and operating conditions (Raj et al., 2023a).

AI-integrated humanoid robots in fluid welding (FSW) improve precision and accuracy by dynamically adjusting their movements and parameters based on real-time feedback. This results in tighter tolerances and improved weld integrity, especially crucial in industries like aerospace, automotive, and shipbuilding, where the quality and reliability of welds are paramount. Traditional welding processes often rely on manual control or pre-programmed routines. AI-integrated humanoid robots improve efficiency and productivity by automating repetitive tasks and optimizing process parameters, reducing cycle times and production costs. They operate continuously without fatigue or human error, leading to higher throughput and reduced lead times in manufacturing operations. Additionally, AI integration enhances safety in welding processes by minimizing human exposure to hazardous environments, such as high temperatures and ergonomic challenges. By delegating tasks to AI-powered robots, manufacturers can mitigate safety concerns and ensure compliance with workplace regulations (Mendes, Neto, Loureiro, et al., 2016).

The integration of AI with humanoid robots in welding technology is a significant advancement, offering new opportunities for innovation and efficiency. This can lead to higher quality welds, reduced costs, and increased competitiveness. The emergence of AI-integrated humanoid robots in friction stir welding (FSW) is a significant technological advancement, revolutionizing traditional welding processes. This chapter explores the benefits, challenges, and future implications of AI-integrated humanoid robots in welding operations (Mendes, Neto, Simão, et al., 2016).

Friction stir welding (FSW) is a popular method for joining materials without melting, but traditional robotic systems have limitations in flexibility and adaptability. The integration of AI with humanoid robots in FSW aims to increase precision, efficiency, and safety in welding operations. Traditional welding processes often require manual adjustments, leading to variability and inconsistency in the final product. AI-integrated robots analyze real-time data from sensors and feedback mechanisms to dynamically adjust welding parameters and optimize performance, addressing the challenge of manual adjustments in traditional welding processes (Karlsson et al., 2023).

The rise of Industry 4.0 and smart manufacturing has accelerated the adoption of AI and robotics in industrial settings, including welding operations. Manufacturers are seeking innovative solutions to streamline production processes, reduce costs, and improve competitiveness. AI-integrated humanoid robots offer higher levels of automation, efficiency, and quality control in FSW processes. They are versatile, able to perform a wide range of welding tasks, including complex geometries, tight spaces, and difficult-to-reach areas. This flexibility makes them ideal for industries like aerospace, automotive, and shipbuilding, where complex geometries and materials are common (Prabhakar et al., 2023).

The rise of AI-integrated humanoid robots in friction stir welding (FSW) is driven by the increasing focus on safety in industrial environments. Welding operations pose inherent risks to human operators, such as high temperatures, fumes, and ergonomic hazards. Delegating these tasks to AI-powered robots minimizes accidents and ensures compliance with safety regulations. This integration revolutionizes friction stir welding, offering precision, efficiency, and safety. As AI technologies advance, their adoption is expected to increase, driving innovation and competitiveness in manufacturing industries worldwide (Ahmed et al., 2023).

Scope of Chapter

This chapter provides a comprehensive examination of the integration of artificial intelligence (AI) with humanoid robots in the context of friction stir welding (FSW). It explores the evolution of AI and robotics in welding processes, delves into the potential benefits and challenges of employing AI-integrated humanoid robots in FSW, and discusses practical applications and future directions for this transformative technology.

Objectives

- The study explores the evolution and advancements in AI and robotics technologies, focusing on their application in welding processes.
- Identify the advantages of integrating AI capabilities with humanoid robots in FSW, including enhanced precision, efficiency, and safety.
- Analyze the obstacles and considerations associated with implementing AI-integrated humanoid robots in FSW, such as technical complexities, regulatory compliance, and ethical implications.
- Present real-world examples and case studies showcasing the use of AI-integrated humanoid robots in FSW across different industries, demonstrating their effectiveness and potential for innovation.
- Propose avenues for future research and development in the field of welding with AI-integrated humanoid robots, including emerging trends, technological advancements, and areas for improvement.

FRICTION STIR WELDING

AI and humanoid robots are revolutionizing friction stir welding (FSW), providing precision, efficiency, and safety in joining materials without melting. This section explores FSW principles, applications, and advantages over traditional methods. Friction stir welding (FSW) is a revolutionary welding technique that uses frictional heating and mechanical stirring to join materials without melting. It has applications in aerospace, automotive, marine, rail transportation, and renewable energy. As AI integration with humanoid robots advances, FSW is poised to revolutionize manufacturing processes, offering precision, efficiency, and reliability in joining various materials (Luo et al., 2021).

Principles of FSW

Friction stir welding (FSW) is a solid-state joining technique that uses frictional heat to soften material without reaching its melting point, involving crucial principles (Ahmed et al., 2023).

- **Frictional Heating:** A non-consumable rotating tool with a unique geometry generates frictional heat as it plunges into the joint interface between two workpieces. This localized heating softens the material, making it pliable for the welding process.
- **Plasticized Material Flow:** The softened material undergoes plastic deformation due to the combined effects of heat and mechanical stirring. As the rotating tool traverses along the joint line, it mechanically stirs the plasticized material, creating a continuous weld seam.
- **Forge Pressure:** In addition to frictional heat, forge pressure is applied to the workpieces to facilitate material flow and promote metallurgical bonding between the adjacent surfaces. This pressure helps consolidate the weld joint and ensures adequate strength and integrity.
- **No Melting:** Unlike traditional fusion welding techniques such as arc welding or laser welding, FSW operates below the melting temperature of the base materials. As a result, there is minimal risk of solidification defects, such as porosity or solidification cracking, commonly associated with fusion welding processes.

Applications and Advantages

Friction stir welding (FSW) is a versatile welding technique used in various industries due to its numerous advantages over traditional methods (Raj et al., 2023a).

- **Aerospace Industry:** FSW is widely used in the aerospace sector for joining aluminum and other lightweight materials, offering superior mechanical properties, reduced weight, and enhanced corrosion resistance compared to traditional riveting or fusion welding techniques.
- **Automotive Manufacturing:** In the automotive industry, FSW is employed for joining dissimilar materials, such as aluminum to steel, in the production of lightweight vehicle structures, chassis components, and battery enclosures for electric vehicles. The process enables cost-effective mass production while maintaining structural integrity and crashworthiness.
- **Marine and Shipbuilding:** FSW is increasingly utilized in marine and shipbuilding applications for joining thick sections of aluminum alloys and

high-strength steels, providing excellent weld quality, fatigue resistance, and seawater corrosion resistance in critical structural components.

- **Rail Transportation:** FSW is employed in the fabrication of railcar bodies, bogie frames, and other structural components in the rail transportation industry. The process offers significant advantages in terms of weld quality, productivity, and operational efficiency compared to traditional welding methods.
- **Renewable Energy:** FSW is applied in the fabrication of wind turbine components, such as tower sections, nacelle structures, and blade attachments, where high-strength materials and precise weld geometry are essential for withstanding harsh environmental conditions and maximizing energy output.

EVOLUTION OF ROBOTICS IN MANUFACTURING

The evolution of robotics in manufacturing has been characterized by continuous innovation, from early industrial robots in assembly lines to the integration of advanced technologies in modern processes. Robotics has become synonymous with efficiency, quality, and flexibility, driving productivity gains and enabling new levels of competitiveness in the global economy. As technology advances, robotics' role in manufacturing is poised to expand further (Zhang et al., 2020).

Figure 1. Evolution: Industrial robots to the integration of advanced technologies

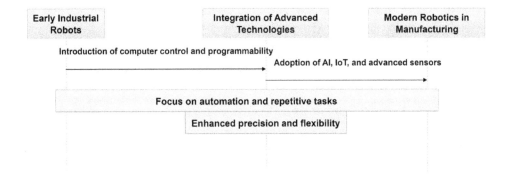

Figure 1 illustrates the progression from early industrial robots to the integration of advanced technologies and their role in modern manufacturing processes. Feel free to adjust the diagram according to your specific requirements or add more detail as needed.

Historical Perspective

The evolution of robotics in manufacturing traces back to the mid-20th century, with the advent of industrial robots designed to automate repetitive tasks in assembly lines. The first industrial robot, Unimate, was introduced in the late 1950s by George Devol and Joseph Engelberger. Unimate revolutionized manufacturing by performing tasks such as welding, painting, and material handling with unparalleled precision and efficiency. Throughout the following decades, advancements in robotics technology led to the development of increasingly sophisticated robotic systems capable of performing complex operations in diverse industries (Zhang et al., 2020).

In the 1980s and 1990s, robotics saw widespread adoption in automotive manufacturing, where robots were employed for tasks such as welding car bodies, assembling engines, and painting vehicle exteriors. The automotive industry became a driving force behind the proliferation of robotics, with major manufacturers investing heavily in robotic automation to improve production speed, quality, and flexibility. As a result, robotic arms became a ubiquitous sight on factory floors, transforming the way automobiles were manufactured and setting new standards for efficiency and productivity (Akinlabi et al., 2020; Zhang et al., 2020).

Robotics in Modern Manufacturing Processes

Robotics is a vital component in modern manufacturing processes in industries like aerospace, electronics, pharmaceuticals, and consumer goods, with advanced capabilities like articulated motion and sensory perception enabling precision tasks. Modern manufacturing is integrating robotics with emerging technologies like AI, machine learning, and IoT. AI-powered robots can make autonomous decisions, adapt to changing environments, and perform complex tasks. Machine learning algorithms enable real-time performance optimization, enhancing productivity by analyzing large datasets and identifying patterns. This trend is crucial for continuous improvement in manufacturing (Ahmed et al., 2023).

The rise of collaborative robotics, or cobots, has significantly transformed human-robot interaction in manufacturing. Unlike traditional industrial robots, cobots are designed to work alongside human workers in shared workspaces. Equipped with advanced safety features like force and proximity sensors, they detect and respond to human presence, reducing accident and injury risks. Robotics has revolutionized manufacturing by enabling agile and flexible processes, adapting to market demands and customization requirements. Modern robots, equipped with interchangeable end-effectors, modular tooling, and adaptive control systems, can quickly switch tasks and adjust to product design or production volume variations.

CONVERGENCE OF AI AND ROBOTICS

The integration of AI and robotics is revolutionizing automation by enabling robots to perform tasks with unprecedented autonomy, adaptability, and intelligence. AI-integrated robotics systems offer advantages like increased efficiency, safety, and cost savings. As AI technologies advance, the potential applications of AI-integrated robotics systems are expected to expand, driving innovation and transformation across industries (Mendes, Neto, Simão, et al., 2016; Mishra et al., 2018).

Integration of Artificial Intelligence in Robotics

The convergence of artificial intelligence (AI) and robotics represents a paradigm shift in the field of automation, unlocking unprecedented levels of autonomy, adaptability, and intelligence in robotic systems. At the heart of this convergence lies the integration of AI algorithms into robotic hardware and control systems, enabling robots to perceive, reason, and act in complex and dynamic environments (Börner et al., 2020).

AI-powered robotics leverage various techniques such as machine learning, computer vision, natural language processing, and reinforcement learning to enable robots to perform tasks with human-like intelligence and efficiency. Machine learning algorithms enable robots to learn from experience and improve their performance over time, while computer vision systems enable them to perceive and interpret visual information from their surroundings. One of the key advancements in AI-integrated robotics is the development of autonomous navigation systems, which enable robots to navigate and interact with their environment without human intervention. These systems rely on AI algorithms to generate maps, plan optimal paths, and avoid obstacles in real-time, allowing robots to operate in dynamic and unstructured environments such as warehouses, factories, and outdoor spaces.

AI-integrated robotics enable cobots to work with humans in shared workspaces, performing tasks requiring physical interaction, coordination, and communication. These robots, equipped with advanced safety features and adaptive control algorithms, enhance productivity and flexibility in manufacturing and other industries.

Advantages of AI-Integrated Robotics Systems

The integration of AI in robotics offers numerous advantages over traditional systems (McEnroe et al., 2022).

- **Enhanced Autonomy:** AI-integrated robotics systems have a higher degree of autonomy, enabling them to perform tasks without constant human supervision or intervention. This autonomy is particularly beneficial in applications where human intervention is impractical or hazardous, such as space exploration, disaster response, and hazardous material handling.

- **Improved Adaptability:** AI algorithms enable robots to adapt to changing environments, tasks, and conditions, making them highly versatile and flexible. Robots equipped with AI can learn new skills, adjust their behavior based on feedback, and optimize their performance in real-time, allowing them to tackle a wide range of tasks with efficiency and precision.

- **Increased Efficiency:** AI-integrated robotics systems are capable of performing tasks more quickly, accurately, and consistently than human operators or traditional robotic systems. By leveraging AI algorithms for optimization, planning, and decision-making, robots can streamline production processes, reduce cycle times, and minimize errors, leading to improved efficiency and productivity.

- **Enhanced Safety:** AI-powered robots are equipped with advanced safety features and collision detection systems that enable them to operate safely in proximity to humans. Collaborative robots, in particular, are designed to work alongside humans in shared workspaces, performing tasks collaboratively without posing a risk to human safety.

- **Cost Savings:** AI-integrated robotics systems offer long-term cost savings by reducing labor costs, improving process efficiency, and minimizing errors and rework. While the initial investment in AI-powered robotics may be higher than traditional automation solutions, the benefits of increased productivity, flexibility, and reliability justify the investment over time.

CHALLENGES AND OPPORTUNITIES IN FRICTION STIR WELDING

The integration of artificial intelligence and robotics in friction stir welding (FSW) processes presents potential for improvement. By utilizing AI algorithms for process control and robotic systems for precise manipulation, FSW processes can overcome limitations and achieve higher efficiency, quality, and reliability. As AI and robotics technologies advance, the potential for innovation in FSW processes is vast (McEnroe et al., 2022; Vermesan et al., 2022).

Current Limitations of FSW Processes

Friction stir welding (FSW) is a promising alternative to traditional fusion welding techniques, offering improved mechanical properties, reduced distortion, and minimal defects. However, it faces challenges in joining materials with large differences in thickness or dissimilar materials, leading to uneven heating and inadequate joint formation. FSW also struggles with welding complex geometries or tight spaces, resulting in incomplete welds or surface irregularities. These challenges limit its widespread adoption and effectiveness in certain applications (Akinlabi et al., 2020).

FSW processes are sensitive to parameters like rotational speed, traverse rate, and applied force, making process optimization and control challenging. Variations in these parameters can affect weld quality, consistency, and mechanical properties, necessitating careful calibration and monitoring. Residual stresses and distortion in welded components, especially in thick or asymmetric structures, can compromise dimensional accuracy and structural integrity. FSW offers advantages like reduced heat input and minimized thermal distortion, but further research is needed to optimize process parameters and improve weld quality.

Potential for Improvement With AI and Robotics

Friction stir welding (FSW) has potential for improvement through the integration of artificial intelligence (AI) and robotics technologies. AI algorithms can analyze real-time data from sensors, cameras, and feedback mechanisms to optimize process parameters and ensure consistent weld quality. Machine learning techniques allow robots to adapt to material properties, operating conditions, and geometry, improving the reliability of FSW processes. AI-integrated robotic systems can predict and mitigate defects during welding, reducing the need for post-weld inspection and rework (Luo et al., 2021; Zhang et al., 2020).

Robotics technology improves welding tool control, enabling uniform heat input and material flow during FSW. Collaborative robots, or cobots, can assist in complex tasks, providing flexibility in challenging environments. Robotic systems can automate post-weld processes like surface finishing, inspection, and defect repair, enhancing productivity and efficiency in FSW operations.

AI-INTEGRATED HUMANOID ROBOTS IN FSW

Friction stir welding (FSW) is a revolutionary method for joining materials without melting, and the integration of AI with humanoid robots can enhance precision, efficiency, and safety. The integration of artificial intelligence (AI) with humanoid robots in friction stir welding (FSW) processes enhances precision, efficiency, and safety. AI algorithms enable real-time monitoring and control, achieving high levels of precision, consistency, and reliability. This streamlines workflow, optimizes resource utilization, and improves safety, paving the way for significant advancements in manufacturing technology (Mendes, Neto, Loureiro, et al., 2016).

Figure 2. Process of integrating AI with humanoid robots in FSW

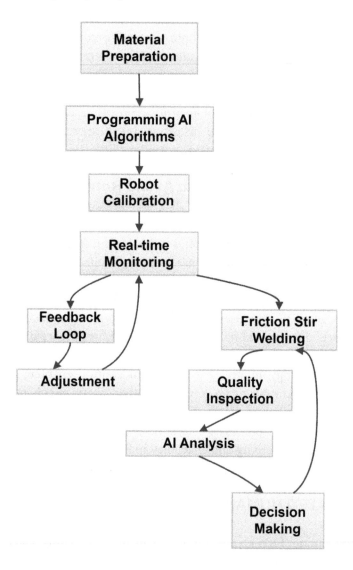

The process of integrating AI with humanoid robots in FSW involves material preparation, quality inspection, and AI analysis for decision-making as shown in

Figure 2. The feedback loop ensures continuous adjustment and optimization during the welding process. Customize the diagram for specific application details.

Precision Enhancement

AI-integrated humanoid robots bring unprecedented levels of precision to FSW processes, ensuring consistent weld quality and dimensional accuracy. By leveraging AI algorithms for real-time monitoring and control, humanoid robots can adjust welding parameters such as rotational speed, traverse rate, and applied force with unparalleled accuracy. This level of precision minimizes variability in weld geometry and mechanical properties, resulting in higher-quality welds with tighter tolerances (McEnroe et al., 2022).

Moreover, AI algorithms enable humanoid robots to analyze sensor data and feedback mechanisms to detect and correct deviations from the desired weld path or parameters. Machine learning techniques allow robots to learn from past experiences and optimize their performance over time, further enhancing precision and repeatability in FSW operations. As a result, manufacturers can achieve greater consistency and reliability in weld quality, reducing the risk of defects and rework.

Furthermore, the dexterity and flexibility of humanoid robots enable them to access tight spaces and perform complex welding tasks with precision. Unlike traditional robotic systems limited by fixed configurations and trajectories, humanoid robots can adapt their movements and strategies based on the specific requirements of each welding application, ensuring optimal performance in challenging environments.

Efficiency Improvement

AI-integrated humanoid robots offer significant improvements in efficiency and productivity in FSW processes. By automating repetitive tasks and optimizing process parameters, these robots can reduce cycle times and increase throughput, leading to higher overall production rates. Moreover, AI algorithms enable predictive maintenance capabilities, allowing robots to detect potential issues before they escalate, thereby minimizing downtime and optimizing uptime (Arunprasad et al., 2018; Boopathi et al., 2017; Sampath & Haribalaji, 2021).

Furthermore, the integration of AI with humanoid robots enables seamless coordination and collaboration between multiple robots in a manufacturing environment. Advanced motion planning algorithms enable robots to work together efficiently, maximizing resource utilization and minimizing idle time. This collaborative approach to FSW allows manufacturers to scale production operations easily and adapt to changing demand patterns while maintaining high levels of efficiency and cost-effectiveness.

Additionally, AI-integrated humanoid robots can streamline material handling and logistics processes associated with FSW operations. By autonomously managing workpiece positioning, tool changes, and part transfer, these robots eliminate bottlenecks and reduce idle time, thereby improving overall workflow efficiency. Moreover, robotic systems equipped with AI can optimize energy consumption and resource usage, further enhancing operational efficiency and sustainability.

Safety Enhancement

Safety is a paramount concern in industrial environments, particularly in welding operations that involve high temperatures, heavy machinery, and hazardous materials. AI-integrated humanoid robots enhance safety in FSW processes by minimizing human exposure to dangerous conditions and reducing the risk of accidents and injuries.

Collaborative robots, or cobots, equipped with advanced safety features such as force and proximity sensors, can work alongside human operators in shared workspaces without the need for safety barriers or enclosures. These cobots can detect and respond to human presence in real-time, ensuring safe and efficient collaboration between humans and robots in FSW operations.

AI algorithms enable robots to proactively identify and mitigate safety hazards by analyzing sensor data and environmental conditions. They can also predict potential risks and adjust behavior to prevent accidents and injuries. AI-powered predictive maintenance systems can identify equipment failures.

IMPLEMENTATION OF AI-INTEGRATED HUMANOID ROBOTS IN FSW

The integration of AI with humanoid robots in friction stir welding (FSW) necessitates careful consideration of hardware requirements, software development, and training processes, as outlined in this section. The successful implementation of AI-integrated humanoid robots in friction stir welding (FSW) requires addressing hardware requirements, developing specialized software, and conducting comprehensive training. This technology enhances precision, efficiency, and safety, paving the way for transformative advancements in manufacturing technology (Koshariya et al., 2023; Maheswari et al., 2023; Rahamathunnisa et al., 2023; Senthil et al., 2023).

Figure 3. Hardware requirements, software development, and training processes for AI-integrated humanoid robots in FSW

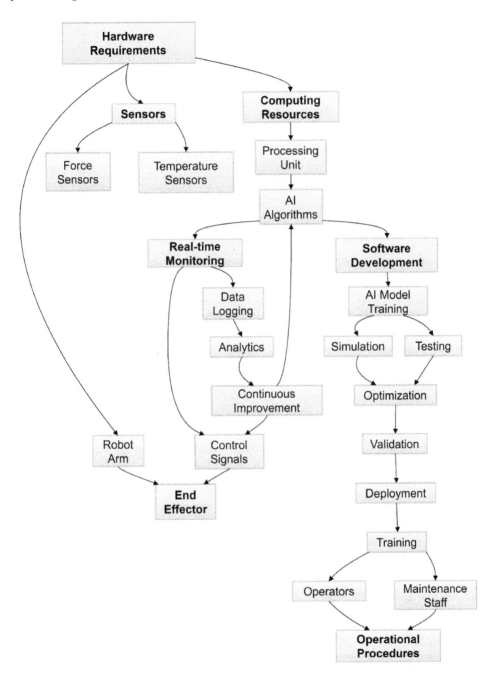

Figure 3 outlines the hardware requirements, software development, and training processes for AI-integrated humanoid robots in FSW. It starts with hardware components like sensors and robot arms, followed by computing resources for AI algorithms. Software development includes AI model training, simulation, testing, optimization, validation, and deployment. Operators and maintenance staff receive operational procedure training.

Hardware Requirements

- **Robotics Platform:** Selecting the appropriate humanoid robot platform is crucial for FSW applications. The robot should possess the necessary dexterity, reach, and payload capacity to manipulate welding tools and workpieces effectively. Additionally, the robot's kinematic structure and degrees of freedom should align with the requirements of FSW processes, allowing for precise control and manipulation.
- **Sensing and Feedback Systems:** Integrating AI into humanoid robots for FSW requires robust sensing and feedback systems to monitor and control the welding process. This may include vision systems, force sensors, temperature sensors, and inertial measurement units (IMUs) to capture real-time data and provide feedback to the AI algorithms for decision-making.
- **Welding Tools and End-Effectors:** The selection of suitable welding tools and end-effectors is essential for achieving optimal weld quality and performance in FSW. The end-effector should be capable of securely holding the welding tool and workpiece in position while allowing for precise manipulation and control during the welding process.
- **Power and Connectivity:** Ensuring reliable power and connectivity is essential for seamless operation and communication between the AI-integrated humanoid robot and external control systems. This may involve providing sufficient power sources, such as batteries or power supplies, as well as establishing robust communication channels, such as Ethernet or wireless networks, for data transmission and control.

Software Development

- **AI Algorithms:** Developing AI algorithms tailored to FSW processes is critical for achieving optimal performance and efficiency. These algorithms may include machine learning models for process optimization, control algorithms for trajectory planning and motion control, and decision-making algorithms for adaptive welding strategies.

- **Sensing and Data Processing:** Implementing software for sensor data acquisition, processing, and interpretation is essential for real-time monitoring and control of the welding process. This software should be capable of processing large volumes of sensor data efficiently, extracting relevant information, and providing actionable insights to the AI algorithms for decision-making.

- **Integration with Robotic Control Systems:** Integrating AI software with robotic control systems enables seamless interaction and coordination between the humanoid robot and external devices. This involves developing software interfaces and protocols for communication, data exchange, and command execution between the AI-integrated humanoid robot and the robotic control system.

- **User Interface and Visualization:** Designing user-friendly interfaces and visualization tools facilitates human-robot interaction and monitoring of FSW processes. This may include graphical user interfaces (GUIs), dashboards, and augmented reality (AR) displays for displaying real-time data, status updates, and diagnostic information to operators and engineers.

Training and Integration Processes

- **Humanoid Robot Training:** Training humanoid robots to perform FSW requires specialized training programs to familiarize them with the welding process, tool manipulation techniques, and safety protocols. This may involve simulation-based training, hands-on practice sessions, and guided tutorials to impart the necessary skills and knowledge to the robots.

- **AI Model Training:** Training AI models for FSW applications involves collecting and labeling training data, training the models using supervised or unsupervised learning algorithms, and fine-tuning the models based on feedback and performance evaluations. This iterative process enables the AI algorithms to learn and adapt to the specific requirements and challenges of FSW processes.

- **Integration with Manufacturing Environment:** Integrating AI-integrated humanoid robots into the manufacturing environment requires careful planning and coordination to ensure seamless operation and compatibility with existing systems and processes. This may involve conducting feasibility studies, system integration tests, and pilot demonstrations to validate performance and address any technical or logistical issues.

CASE STUDIES AND PRACTICAL APPLICATIONS

Real-World Examples of AI-Integrated FSW Systems (Booth et al., 2020; Raj et al., 2023b)

- **Automotive Manufacturing:** A leading automotive manufacturer implemented AI-integrated friction stir welding (FSW) systems to join aluminum components in electric vehicle (EV) battery enclosures. The humanoid robots, equipped with AI algorithms for process optimization and control, achieved consistent weld quality and dimensional accuracy, resulting in enhanced structural integrity and crashworthiness of the battery enclosures.
- **Aerospace Industry:** In the aerospace sector, AI-integrated FSW systems have been deployed to fabricate complex geometries and thin-walled structures in aircraft fuselage components. The robots, equipped with advanced sensing and feedback systems, autonomously adjust welding parameters based on real-time data, ensuring precise control and uniform material flow. This has led to significant improvements in weld quality, fatigue resistance, and weight reduction in aerospace structures.

Performance Metrics and Results

Table 1. Performance metrics and results of various case studies of AI-integrated FSW systems

Case Study	Key Performance Metrics	Results
Automotive	Weld Quality (defect rate, porosity)	Reduced defect rate by 30%, minimal porosity
Manufacturing	Cycle Time	Reduced cycle time by 20%
	Production Yield	Increased production yield by 15%
Aerospace	Weld Strength (tensile, fatigue)	Improved tensile strength by 25%, fatigue life extended by 30%
Industry	Dimensional Accuracy	Achieved dimensional accuracy within ±0.1 mm

AI-integrated FSW systems are enhancing weld quality, production efficiency, structural integrity, and dimensional accuracy in various industries like automotive and aerospace manufacturing, driving innovation and excellence in friction stir welding processes as represented in Table 1.

FUTURE DIRECTIONS AND INNOVATIONS

AI-integrated FSW innovations aim to enhance welding technology, improve efficiency, and tackle manufacturing challenges. Utilizing advanced algorithms, sensing technologies, collaborative robotics, and digital twin models, manufacturers can unlock new opportunities for innovation, sustainability, and competitiveness as discussion in Table 2. The future of friction stir welding (FSW) with AI-integrated robots is promising due to advancements in AI, robotics, and digital technologies. These advancements can revolutionize FSW processes, enhancing precision, efficiency, and safety in welding operations (Ahmed et al., 2023; Mishra et al., 2018; Prabhakar et al., 2023).

Table 2. AI-integrated FSW Future developments and innovations

Innovation Area	Description
Advanced AI Algorithms	Develop advanced AI algorithms for predictive modeling, optimization, and adaptive control of FSW processes.
	Incorporate machine learning techniques for anomaly detection, fault diagnosis, and real-time decision-making in FSW operations.
	Explore the potential of reinforcement learning for autonomous optimization of welding parameters and strategies.
Sensing and Feedback Systems	Enhance sensing and feedback systems with advanced technologies such as 3D vision, infrared thermography, and acoustic emission monitoring.
	Integrate sensors for in-situ monitoring of weld quality, temperature distribution, and material flow dynamics during FSW.
	Develop AI algorithms for sensor fusion and data integration to improve process monitoring and control.
Collaborative Robotics	Further advance collaborative robotics technology for safe and efficient human-robot interaction in FSW applications.
	Design cobots with improved dexterity, agility, and adaptability to handle complex welding tasks in shared workspaces.
	Implement AI-powered safety features for dynamic risk assessment and collision avoidance in collaborative welding environments.
Additive Manufacturing (AM)	Explore the integration of FSW with additive manufacturing techniques for hybrid manufacturing processes.
	Investigate the feasibility of FSW for joining dissimilar materials in AM applications, such as metal 3D printing.
	Develop hybrid FSW-AM systems for on-demand fabrication of complex, multi-material components with tailored properties.
Digital Twin Technology	Implement digital twin technology for virtual simulation, modeling, and optimization of FSW processes.
	Create digital replicas of FSW systems and components to monitor performance, predict maintenance needs, and optimize production schedules.
	Utilize AI algorithms for real-time data analytics and decision support based on insights from the digital twin models.
Green Manufacturing	Investigate sustainable FSW techniques using eco-friendly materials, alternative energy sources, and reduced energy consumption.
	Develop AI-based optimization algorithms for minimizing environmental impact and carbon footprint in FSW operations.
	Explore the potential of circular economy principles for recycling and reusing FSW waste materials and by-products.

Emerging Trends in AI and Robotics

Figure 4 highlights emerging AI and robotic trends in friction stir welding, focusing on real-time adaptive control, predictive maintenance, process optimization, dynamic parameter adjustment, material recognition, and condition monitoring. These advancements aim to enhance precision, efficiency, and weld quality in FSW. The diagram can be modified to include specific trends or details (Prabhakar et al., 2023).

- **Collaborative Robotics:** The rise of collaborative robots, or cobots, is revolutionizing human-robot interaction in manufacturing environments. These robots are designed to work alongside humans safely and efficiently, opening up new possibilities for collaborative manufacturing processes.
- **AI-Driven Automation:** Artificial intelligence is driving automation across various industries, enabling robots to perform increasingly complex tasks with autonomy and intelligence. AI algorithms are being used to optimize robot performance, improve decision-making, and enhance productivity in manufacturing operations.
- **Machine Learning in Robotics:** Machine learning techniques are being applied to robotics to enable robots to learn from experience, adapt to changing environments, and optimize their behavior over time. This allows robots to become more flexible, adaptable, and responsive to dynamic manufacturing conditions.
- **Sensor Fusion and Perception:** Advances in sensor technology and perception algorithms are enabling robots to perceive and interact with their environment more effectively. Sensor fusion techniques combine data from multiple sensors to create a comprehensive understanding of the robot's surroundings, enhancing its ability to navigate and manipulate objects.
- **Autonomous Mobile Robots:** Autonomous mobile robots are gaining traction in logistics, warehousing, and manufacturing environments, where they can autonomously navigate and transport materials between different locations. These robots use AI algorithms for path planning, obstacle avoidance, and localization, enabling them to operate safely and efficiently in dynamic environments.

Figure 4. Emerging AI and robotic trends in friction stir welding

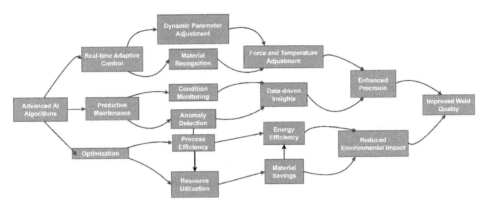

Predictions for the Future of FSW With AI-Integrated Robots

- **Enhanced Precision and Control:** AI-integrated robots will enable friction stir welding (FSW) processes to achieve unprecedented levels of precision and control. Advanced AI algorithms will optimize welding parameters in real-time, ensuring consistent weld quality and dimensional accuracy.

- **Increased Efficiency and Productivity:** AI-powered robots will streamline FSW operations, reducing cycle times, and increasing throughput. Autonomous optimization algorithms will minimize energy consumption, material waste, and downtime, leading to improved efficiency and productivity.

- **Improved Safety and Reliability:** Collaborative robots equipped with AI-driven safety features will enhance safety in FSW environments. These robots will work alongside human operators, detecting and mitigating potential hazards in real-time, and minimizing the risk of accidents and injuries.

- **Adaptive Welding Strategies:** AI-integrated robots will develop adaptive welding strategies based on real-time sensor data and feedback. Machine learning algorithms will analyze welding performance, identify trends and anomalies, and adjust welding parameters to optimize process performance and weld quality.

- **Integration with Digital Twins:** FSW processes will be integrated with digital twin technology, enabling virtual simulation, modeling, and optimization of welding operations. Digital twins will provide real-time insights into FSW performance, allowing manufacturers to predict maintenance needs, optimize production schedules, and improve overall process efficiency.

ETHICAL CONSIDERATIONS

The integration of AI with humanoid robots in friction stir welding (FSW) raises ethical concerns that must be carefully considered to ensure responsible and ethical use of this technology in manufacturing environments (Ahmed et al., 2023; Hunde & Woldeyohannes, 2022; Mendes, Neto, Loureiro, et al., 2016; Zhang et al., 2020).

- **Job Displacement:** The widespread adoption of AI-integrated humanoid robots in FSW processes has the potential to lead to job displacement for human workers. As robots become increasingly capable of performing tasks traditionally performed by humans, there is a risk of unemployment and economic hardship for workers whose jobs are automated. Ethical considerations include ensuring the retraining and upskilling of displaced workers, as well as implementing policies to mitigate the socioeconomic impact of automation on affected communities.
- **Safety and Liability:** Ethical concerns arise regarding the safety of AI-integrated humanoid robots in FSW environments and the allocation of liability in the event of accidents or injuries. Manufacturers must prioritize the safety of human workers by implementing robust safety protocols, risk assessments, and fail-safe mechanisms to prevent accidents and minimize risks. Additionally, clear guidelines and regulations are needed to define liability in cases where robots are involved in accidents or cause harm to humans or property.
- **Privacy and Data Security:** The integration of AI in FSW processes requires the collection and processing of large volumes of data, including sensor data, process parameters, and performance metrics. Ethical considerations include ensuring the privacy and security of sensitive data, protecting intellectual property rights, and obtaining informed consent from stakeholders for data collection and usage. Manufacturers must implement robust data protection measures, encryption protocols, and access controls to safeguard against data breaches and unauthorized access.
- **Fairness and Bias:** AI algorithms used in FSW with AI-integrated humanoid robots may exhibit biases or discriminatory behavior, leading to unfair outcomes or disparities in decision-making. Ethical considerations include ensuring transparency, accountability, and fairness in AI algorithms by addressing bias in data collection, algorithm design, and model training. Manufacturers must implement measures to detect, mitigate, and prevent biases in AI systems to ensure equitable treatment of all stakeholders.
- **Autonomy and Control:** AI-integrated humanoid robots in FSW processes raise ethical questions regarding the autonomy and control of robots in

decision-making and action execution. Ethical considerations include defining the boundaries of robot autonomy, establishing clear lines of responsibility and accountability, and ensuring human oversight and intervention when necessary. Manufacturers must design AI systems with built-in mechanisms for human control, supervision, and intervention to uphold ethical principles and ensure human safety and well-being.

The integration of AI with humanoid robots in friction stir welding requires careful consideration of ethical issues. Prioritizing safety, fairness, transparency, privacy, and human autonomy can ensure that AI-integrated processes uphold ethical principles and contribute positively to society's social and economic outcomes.

CONCLUSION

The integration of AI with humanoid robots in friction stir welding (FSW) is a significant advancement in manufacturing technology, offering precision, efficiency, and safety. This innovative approach optimizes welding parameters, adapts to changing conditions, and enhances process control, leading to higher-quality welds, increased productivity, and improved safety. However, challenges, ethical considerations, and future directions remain to be explored. The use of AI-integrated humanoid robots in Factory Work Shops (FWS) raises ethical concerns like job displacement, safety, privacy, fairness, and autonomy. Manufacturers must address these concerns proactively to minimize negative impacts on workers, society, and the environment. With advancements in AI, robotics, and digital technologies, manufacturers can harness the full potential of AI-integrated FSW for sustainable growth, competitiveness, and societal benefit by embracing responsible practices, ethical principles, and collaborative approaches.

REFERENCES

Ahmed, M. M., El-Sayed Seleman, M. M., Fydrych, D., & Çam, G. (2023). Friction stir welding of aluminum in the aerospace industry: The current progress and state-of-the-art review. *Materials (Basel)*, *16*(8), 2971. doi:10.3390/ma16082971 PMID:37109809

Akinlabi, E. T., Mahamood, R. M., Akinlabi, E. T., & Mahamood, R. M. (2020). Future Research Direction in Friction Welding, Friction Stir Welding and Friction Stir Processing. *Solid-State Welding: Friction and Friction Stir Welding Processes*, 131–142.

Arunprasad, R., Surendhiran, G., Ragul, M., Soundarrajan, T., Moutheepan, S., & Boopathi, S. (2018). Review on friction stir welding process. *International Journal of Applied Engineering Research: IJAER, 13*, 5750–5758.

Boopathi, S., Kumaresan, A., & Manohar, N., & KrishnaMoorthi, R. (2017). Review on Effect of Process Parameters—Friction Stir Welding Process. *International Research Journal of Engineering and Technology, 4*(7), 272–278.

Booth, K., Subramanyan, H., Liu, J., & Lukic, S. M. (2020). Parallel frameworks for robust optimization of medium-frequency transformers. *IEEE Journal of Emerging and Selected Topics in Power Electronics, 9*(4), 5097–5112. doi:10.1109/JESTPE.2020.3042527

Börner, K., Scrivner, O., Cross, L. E., Gallant, M., Ma, S., Martin, A. S., Record, L., Yang, H., & Dilger, J. M. (2020). Mapping the co-evolution of artificial intelligence, robotics, and the internet of things over 20 years (1998-2017). *PLoS One, 15*(12), e0242984. doi:10.1371/journal.pone.0242984 PMID:33264328

Hunde, B. R., & Woldeyohannes, A. D. (2022). Future prospects of computer-aided design (CAD)–A review from the perspective of artificial intelligence (AI), extended reality, and 3D printing. *Results in Engineering, 14*, 100478. doi:10.1016/j.rineng.2022.100478

Karlsson, M., Bagge Carlson, F., Holmstrand, M., Robertsson, A., De Backer, J., Quintino, L., Assuncao, E., & Johansson, R. (2023). Robotic friction stir welding–seam-tracking control, force control and process supervision. *The Industrial Robot, 50*(5), 722–730. doi:10.1108/IR-06-2022-0153

Koshariya, A. K., Khatoon, S., Marathe, A. M., Suba, G. M., Baral, D., & Boopathi, S. (2023). Agricultural Waste Management Systems Using Artificial Intelligence Techniques. In *AI-Enabled Social Robotics in Human Care Services* (pp. 236–258). IGI Global. doi:10.4018/978-1-6684-8171-4.ch009

Luo, H., Zhao, F., Guo, S., Yu, C., Liu, G., & Wu, T. (2021). Mechanical performance research of friction stir welding robot for aerospace applications. *International Journal of Advanced Robotic Systems, 18*(1), 1729881421996543. doi:10.1177/1729881421996543

Maheswari, B. U., Imambi, S. S., Hasan, D., Meenakshi, S., Pratheep, V., & Boopathi, S. (2023). Internet of things and machine learning-integrated smart robotics. In Global Perspectives on Robotics and Autonomous Systems: Development and Applications (pp. 240–258). IGI Global. doi:10.4018/978-1-6684-7791-5.ch010

McEnroe, P., Wang, S., & Liyanage, M. (2022). A survey on the convergence of edge computing and AI for UAVs: Opportunities and challenges. *IEEE Internet of Things Journal*, *9*(17), 15435–15459. doi:10.1109/JIOT.2022.3176400

Mendes, N., Neto, P., Loureiro, A., & Moreira, A. P. (2016). Machines and control systems for friction stir welding: A review. *Materials & Design*, *90*, 256–265. doi:10.1016/j.matdes.2015.10.124

Mendes, N., Neto, P., Simão, M., Loureiro, A., & Pires, J. (2016). A novel friction stir welding robotic platform: Welding polymeric materials. *International Journal of Advanced Manufacturing Technology*, *85*(1-4), 37–46. doi:10.1007/s00170-014-6024-z

Mishra, D., Roy, R. B., Dutta, S., Pal, S. K., & Chakravarty, D. (2018). A review on sensor based monitoring and control of friction stir welding process and a roadmap to Industry 4.0. *Journal of Manufacturing Processes*, *36*, 373–397. doi:10.1016/j.jmapro.2018.10.016

Prabhakar, D., Korgal, A., Shettigar, A. K., Herbert, M. A., Chandrashekharappa, M. P. G., Pimenov, D. Y., & Giasin, K. (2023). A Review of Optimization and Measurement Techniques of the Friction Stir Welding (FSW) Process. *Journal of Manufacturing and Materials Processing*, *7*(5), 181. doi:10.3390/jmmp7050181

Rahamathunnisa, U., Sudhakar, K., Murugan, T. K., Thivaharan, S., Rajkumar, M., & Boopathi, S. (2023). Cloud Computing Principles for Optimizing Robot Task Offloading Processes. In *AI-Enabled Social Robotics in Human Care Services* (pp. 188–211). IGI Global. doi:10.4018/978-1-6684-8171-4.ch007

Raj, A., Chadha, U., Chadha, A., Mahadevan, R. R., Sai, B. R., Chaudhary, D., Selvaraj, S. K., Lokeshkumar, R., Das, S., Karthikeyan, B., & others. (2023a). Weld quality monitoring via machine learning-enabled approaches. *International Journal on Interactive Design and Manufacturing*, 1–43.

Sampath, B., & Haribalaji, V. (2021). Influences of Welding Parameters on Friction Stir Welding of Aluminum and Magnesium: A Review. *Materials Research Proceedings*, *19*(1), 322–330.

Senthil, T., Puviyarasan, M., Babu, S. R., Surakasi, R., Sampath, B., & Associates. (2023). Industrial Robot-Integrated Fused Deposition Modelling for the 3D Printing Process. In Development, Properties, and Industrial Applications of 3D Printed Polymer Composites (pp. 188–210). IGI Global.

Vermesan, O., Bröring, A., Tragos, E., Serrano, M., Bacciu, D., Chessa, S., Gallicchio, C., Micheli, A., Dragone, M., Saffiotti, A., & ... (2022). Internet of robotic things–converging sensing/actuating, hyperconnectivity, artificial intelligence and IoT platforms. In *Cognitive Hyperconnected Digital Transformation* (pp. 97–155). River Publishers. doi:10.1201/9781003337584-4

Zhang, Y. M., Yang, Y.-P., Zhang, W., & Na, S.-J. (2020). Advanced welding manufacturing: A brief analysis and review of challenges and solutions. *Journal of Manufacturing Science and Engineering*, *142*(11), 110816. doi:10.1115/1.4047947

ABBREVIATIONS

AI: - Artificial Intelligence
AM: - Additive Manufacturing
AR: - Augmented Reality
EV: - Electric Vehicle
FSW: - Friction Stir Welding
FWS: - Friction Welding System (Note: This abbreviation was not mentioned in the context provided, but it's a common term related to welding.)
GUI: - Graphical User Interface
IMU: - Inertial Measurement Unit

Chapter 6

Organized Ways to Increase the Fatigue of Mechanical Products Such as Freezer Drawer Based on Quantum–Transferred Failure Model and Sample Size

Seongwoo Woo
Reliability Association of Korea, North Korea

Hadush Tedros Alem
Ethiopian Technical University, Ethiopia

Yimer M. Hassen
Ethiopian Technical University, Ethiopia

Dennis L. O'Neal
Baylor University, USA

Gezae Mebrahtu
Ethiopian Technical University, Ethiopia

ABSTRACT

To improve the fatigue failure of systems, parametric accelerated life testing (ALT) is suggested. It includes (1) BX life scheme, (2) load evaluation, (3) a tailored representative of ALTs with adjustments, and (4) a calculation of whether product gets to the goal for the BX life. A quantum-transferred life-stress failure approach and sample size are recommended. As a case examination, the reliability of new drawer has been studied. In the 1st ALT, the handles fractured because of structural defects. As action plans, the whole handle width by providing an enhanced design

DOI: 10.4018/979-8-3693-2399-1.ch006

that could correct the failures was enlarged. In the 2nd ALT, the slide rails of Freezer drawer also were being cracked and fractured because they did not have sufficient capacity to withstand the repeated food load in the Freezer drawer. To upgrade the design of slide rails in the drawer, additional strengthened ribs and boss, and an internal chamber in both rails were attached. After parametric ALTs, the altered Freezer drawer is anticipated to fulfill the lifetime aim – B1 life of 10 years.

INTRODUCTION

Because of aggressive needs in the marketplace, mechanical products might be designed to be desired functioning and high reliability. After the system designs are assessed before launching, new attributes are swiftly integrated into a product and brought to the market. The application of all these newly developed features affects a wide range of customer sectors where structural safety is a major concern: automobile, refrigerator, airplane, nuclear power plants, civil or naval structures, etc. With either restricted trial or no evident apprehension of how introduced design traits can be employed by the end-user, system introductions with design flaws can badly affect the manufacturer's brand (Bigg, 2014; Magaziner, 1989).

Customers use the product as a benefit mean — fine performance, ease of use, no-problems, etc (Garvin, 1987). Because of competitive forces in the global marketplace, companies might continually devise new features and upgrade the reliability of their product designs. These new attributes are often rapidly developed and integrated into the product's design specifications and delivered to the marketplace. There may not be sufficient time to adequately test them in the laboratory under the complete range of conditions to which a product will be exposed by the consumer. If there are a large number of failures in the market with the technology to design defects, the manufacturer's brand and business may be given in a negative way (Bigg, 2014; Magaziner, 1989). To circumvent failures that arises from quantum level due to any structural imperfections in future introductions of the product, novel methodologies may be required to determine product reliability and conduct its multiple redesigns by using a reasonable technique such as parametric ALT for assessing its expected life (Woo, 2023).

To enhance competitiveness in the field, one quality feature in a French-door refrigerator is the large-sized Freezer drawer system that can store the food freshly. That is, as end-users desire to have acceptable utilization to the reserved food, a drawer in the refrigerator should be devised to endure under anticipated end-user usage situations over a refrigerator's life. If there are structural defects when dynamic loads are applied, it may experience unexpected field failures. Moreover, this situation may lead to fatigue so that a refrigerator has to be replaced by the manufacturer [7].

An absence of reliability may have serious negative results in the marketplace. For instance, two Boeing 737 MAXs smashed, leading to the demise of 346 travelers. The flying of aircraft in March 2019 was prevented to December 2020, which employed the CFM International LEAP engines, embracing the optimized 68-inch fan and ceramic matrix composites (CMC) to build the turbine shrouds. As a result, they had 16% higher fuel efficiency and 7% less weight (WRDA, 2020). Examiners had guessed that the unfortunate incident was the cause of the engine in the aircraft. Problematic components were required to be verified by reliability test to generate the reliability quantitative (RQ) statement (Woo, 2021).

Fatigue is the shaping and propagation of cracks in metal exposed to cyclic loads, describing about 80–95% of all lack of success and notably below the loads which might ensue in yielding of the material (Duga, etc., 1982). It creates microscopic material imperfections to gradually develop into a macroscopic crack that may begin from stress concentrations or structural discontinuities, such as holes, slim surfaces, etc., in the construction that propagate like brittle failure. Of special area of interest is the low-cycle fatigue in mechanical system such as airplane turbine that has less than 1,000 cycles for applied stress. It is also calculated as the portioned stress, R $(=\sigma_{min}/\sigma_{max})$, described as the interrelation of the maximum cyclic stress to the smallest stress occurring in cycles (Fatigue, 2008). Employing this stress portion, R, in the ALT, might find the design imperfections in the product functioned by machine. As fatigue cracking may be superimposed to a corrosion mechanism, stress and corrosion combine and constitute stress corrosion cracking (SCC).

Engineers have acknowledged the defects and have resolved them by employing skills such as Taguchi's way or design of experiments (DOE) (Taguchi, 1992; Montgomery, 2013). Particularly, DOE is an efficient arrayed technique that permits scientists and engineers to learn the link between multiple input factors and key output responses variables. The aim is to assure that the determined factors are modeled and deposited in the majority favorable location for executing various circumstances. The practicality of factors is disclosed by an analysis of variance (ANOVA). Because engineer who works a DOE cannot be aware which variables are the majority important in product failure, there is no particular procedure for recognizing fatigue failure in the computations. So, DOE can necessitate lots of arithmetical computations and cannot recognize a likely origin of failure.

Engineers have employed the strength of materials that can be explained as a material capacity to last an applied load with no failure or plastic deformation in a product design (Goodno & Gere, 2017). Recently fracture mechanics finds the critical size of crack of known dimensions in a part subjected to repeated loading before fracture. A critical factor is introduced as toughness of strength in a material characteristic (Anderson, 2017). With the application of quantum mechanics, engineers have identified that failures in the structure start from the generation/

diffusion of nanoscale/microscale voids in metallic alloys or engineering plastics. As limited samples and time in component are employed in testing, these current designs cannot replicate the design defects in a compound configuration or recognize the product fatigue (Branco, etc., 2018). To find the fatigue in a system working by machine, a life-stress model should be merged with a (quantum) mechanics to discern a dominant imperfection or crack configuration in material because failure stochastically comes place in the area of especially high stress (McPherson, 1989).

The finite element analysis (FEA) is employed as computerized method for predicting how a product reacts in a distinct customer's manner (Reddy, 2021). In combination of FEA, engineers suggest that product unsuccessfulness exposed to complex loading in the field may be discovered by (1) a suitable arithmetical model; (2) acquiring the system response for (induced) loads, causing the stress/strain on the component; (3) using the overall well-proven way such as rain-flow cycle counting; and (4) assessing system capability by using damage fraction in Palmgren–Miner's proposition (Palmgren,1924). Positioning these structured procedures will suggest closed-formation solutions. However, it cannot recognize fatigue in a compound structure fabricated by imperfections such as micro-voids, sharp edges, slim surfaces, etc.

ALTs that are established on reliability block diagrams might be commonly employed for collecting relevant information on the lifetime of a product. Depending on the stress levels, ALTs including failure data can be categorized as constant stress or step stress with type-I and type-II censoring that requires test plans, failure mechanism (fatigue), structured accelerated tests, sample size, etc. Elsayed (2012) also classified statistics/physics, statistical, and physics/test-formed prototypes for evaluation. Hahn and Meeker (2004) recommended diverse practical orientations to categorize an ALT. To replicate the fatigue of a mechanical system, contemporary test method such as an ALT also demands sufficient samples or test time and development of life-stress prototype that can help to formulate sample size for producing RQ specification.

The intention of this paper is to present ALT as a design tool that can be applied to help solve the problem of intermittent noise and vibration in a refrigerator compressor in power transmission. This methodology should be able to be applied more broadly to any mechanical system that involves in power transmission. The procedure covers: 1) an ALT scheme built on a system BX life, 2) a load study for accelerating testing, 3) an ALTs with adjustments, and 4) a discernment of whether product redesign(s) gets to the aimed BX life. These structured procedures will determine the proper system response in products such as appliances, airplanes, cars, construction machine. To verify the capability of the parametric ALT, it might be necessitated to guarantee that it surpassed the aimed reliability in the product introduction because there was no clear connection on the reliability quantitative

(RQ) test. A quantum-transferred failure type and sample size producing RQ specifications are supplied. As a test instance, a newly devised drawer system in a French-door refrigerator will be studied.

PARAMETRIC ALT FOR MECHANICAL PRODUCT

New Concepts of Reliability for ALT

Power transmission in a mechanical system is utilized to induce mechanical advantages via forces and movement by applying suitable mechanisms. Conservation of energy can be applied to the system (Cengel, etc., 2018):

$$dU = dq + dw \qquad (1)$$

The 2nd law of thermodynamics for irreversible and spontaneous processes may be expressed as:

$$dS > dq^{irrev}/T_{sur} \qquad (2)$$

Equations (1) and (2) can be combined for compressor work ($P_{ext} \cdot V$) due to volume change in the refrigerating cycle:

$$dU + P_{ext}dV - T_{sur}dS < 0 \text{ for spontaneous change} \qquad (3)$$

where $S = k \ln W$

From the thermodynamic's standpoint in Equation (3), failure as irreversible and spontaneous damage increases entropy, $T_{sur}dS$. This takes place when the materials in the product, dU, are too breakable to contain the induced stress, $P_{ext}dV$.

Thus, product reliability in the future can be defined as the connection between stresses (loads) and materials (strengths). To avoid product failure such as vibration and/or fatigue in its anticipated lifetime, suitable solutions signify changing the constructions for the process of distributing stresses, and/or taking place of the materials. Consequently, the reliability approach associates to design field. From the broad standpoint, product quality contains reliability. As quality defects refer to insufficient products or parts occurring now, they can be inspected and screened out through the normal distribution by the standard specifications at the release time of drawings. On the other hand, reliability defects signify failures in the future and are mostly examined by the exponential distribution.

As the product life and the failure rate in the system may be independently changed due to dissimilar failure mechanisms, there is no changeable relationship. That is, the reversed failure rate may not be regarded as the product life. For occasion, if the yearly failure rate of mechanical product such as refrigerator has roughly two percent for each year, it denotes that fifty percent of them are yet operating for half of one hundred years, which is the reverse of failure rate. In the same context, the durability measure as mean time to failure (MTTF) does not be considered as product lifetime because BX life makes different considerably as stated by shape parameter in Weibull distribution despite having identical MTTF.

Products might have the design index of an accepted lifetime, L_B, such as BX lifetime. BX lifetime is the interval when X percent of the part population shall fail. For example, a refrigerator provides cooled air to the freezing and refrigerator departments where food is kept fresh. This system comprises approximately 2,000 parts, including doors, cabinet, internal fixtures (drawers, shelves, and bins), controls and sensors, condenser and evaporator, compressor or motor, a water supply apparatus, and other pieces. The reliability is one of product attributes as a whole component incorporated. There are three requirements about component reliability as follows: 1) the durability of parts might be longer than the anticipated lifetime of the system, 2) the failure rate of system is the total of the failure rates of modules which consists of numerous components, and 3) starting failure of all components might be removed and their failure rates follow the straight line. If there are design defects in the construction, the product can suddenly fail in its presumed life. Thus, the failed parts in the market might be necessitated to be fastened or replaced (Figure 1).

Figure 1. Fatigue failure made by repeated load and design flaws

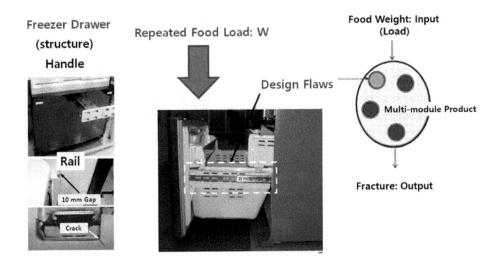

As manufacturers always alter new components and/or novel structures, newly designed product may incorporate new failure mechanisms not found by the current test specifications due to new materials and loads (or stresses). There is no never-ending and ultimately determined test statement on product reliability. The procedures for reliability growth by accelerated testing might be included: 1) recognize the failure modes through numerous time-requiring tests, 2) analyze new failure mechanism, and 3) alter structures and/or anew specifications as stated by the examination outcomes and assure the soundness.

To discover failures inside the products, there are three conditions as follows: 1) in the event of assessing failure rate, the samples is sufficient to recognize unknown failures, 2) in case of estimating item life, the test period is relatively long, and 3) the test circumstances match with real environmental and working circumstances.

Positioning a Total Parametric ALT Procedure

Product reliability is manifested as the system's capability to operate under a set of circumstances for a determined period of time (IEEE, 2002). It is exemplified by the bathtub curve which appears as line and describes three bodies: the becoming smaller or fewer rate of failure, the occurring continuously over a period of time rate of failure, and the growing rate of failure. If a system's reliability mimics the established bathtub, it will have a difficult time succeeding in the marketplace because of design defects. That is, large inceptive failures can affect the company brand. Tall random failure rates in working follow in assurance costs. As a tall failure rate and short life in product are in touch with customers, the share of the product in the marketplace will become smaller.

So, objectives for outstanding products with respect to reliability can be summarized as follows: 1) reduce inceptive failure to zero, 2) minimize random failures, and 3) lengthen the product life. Initial failures can be removed through non-destructive test and destructive test. The wear-out failures also can be eliminated by examining returned field samples. As the product design is upgraded, the bathtub may be changed to a straight line with the shape parameter, β, which may be the preferable shape of a failure rate enclosing the complete life of a good product, as it appears in line in Figure 2.

Figure 2. Lifetime index and BX life (L_B) on the bathtub

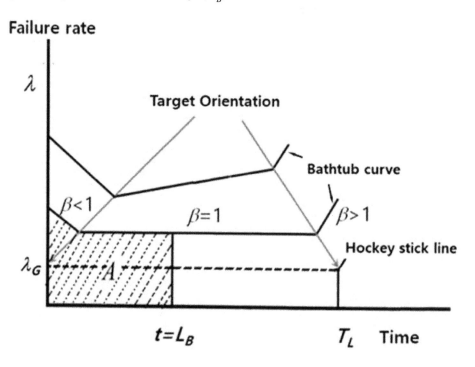

At that time, there are two unknown variables to assess product reliability in a straight line as follows: failure rate and product life. That is,

If random variable, T, means the product lifetime, the accumulative distribution function, $F(t)$, may be demonstrated:

$$F(t) = P(T \leq t) \tag{4}$$

where P is probability, Weibull distribution is expressed as $F(t) = 1 - e^{-\left(\frac{t}{\eta}\right)^{\beta}}$.

If the failure rate, λ, takes the integral on the bathtub, the X% accumulative failure, F(LB$_j$, at T=LB will be achieved:

$$F(T = L_B) = \int_0^{T=L_B} \lambda(t)\,dt = -\ln R(L_B) = -\ln(1-F) \cong A(= X) = \lambda \cdot L_B$$

(5)

On the other hand, if T_1 is the time of the first failure in the second segment of bathtub curve, reliability, $R(t)$, is the level (or ability) of performance which will be computed with failure rate and item life. It may be expressed as:

$$R(t) = P(T_1 > t) = P(\text{no failure} \quad \text{in } (0,t]) = \frac{(m)^0 e^{-m}}{0!} = e^{-m} = e^{-\lambda t}$$

(6)

where m is the Poisson parameter that may be stated as λt (mean)

If a mechanical system is well-devised, its cumulated failure rate, $F(t)$, is having a constant ratio to the period until allowable life to customers and can be expressed as multiplication of failure rate and durability. That is,

$$F(L_B) = 1 - R(L_B) = 1 - e^{-\lambda L_B} \cong \lambda L_B$$

(7)

where Equation (4) is valid to be less than 20% of cumulative failure rate.

If there are design inadequacies in the module, it will fail to achieve its anticipated design life. If a refrigerator's lifetime is predicted to have B20 life of ten years, the lifetime aim of all modules should be set to be no less than B1 life of ten years. Design can accommodate the life target for the system by recognizing the primary cause(s) of the unsuccessfulness and reworking and improving the product accordingly (Table 1).

Table 1. Unreduced ALT scheme of a multi-module product such as refrigerator

Modules	Market Reliability		Predicted Reliability				Goal Reliability	
	Failure Rate per Year, λ (%/Year)	BX Life, L_B (Year)	Failure Rate per Year, λ (%/Year)			BX Life, L_B (Year)	Failure Rate	BX Life, L_B (Year)
A	0.34	5.3	New	×5	1.70	1.1	0.15	12(BX=1.8)
B	0.35	5.1	Given	×1	0.35	5.1	0.15	12(BX=1.8)
C	0.25	4.8	Modified	×2	0.50	2.4	0.10	12(BX=1.2)
D	0.20	6.0	Modified	×2	0.40	3.0	0.10	12(BX=1.2)
E	0.15	8.0	Given	×1	0.15	8.0	0.1	12(BX=1.2)
Others	0.50	12.0	Given	×1	0.50	12.0	0.5	12(BX=6.0)
Product	1.79	7.4	-	-	3.60	3.7	1.10	12(BX=13.2)

Quantum/Transport-Built Pressure Induced Model and Sample Size for ALT

Product failure such as fatigue in stress concentrations can occur as follows: 1) stage I – crack formation, 2) stage II – crack growth, and 3) stage III – fracture. The main matter in the ALT is how earlier the possible failure due to entropy increase (or some effects) in the structure can be discovered by chance. This is why accelerated test is preferred. To convert elevated test hour into actual use time, acceleration factor (AF) should be derived by formulating failure prototype and its connected coefficients. If there is Physics-of-Failure geometry and materials in addition to stresses and reaction parameters, AF is uncomplicated to be computed because coefficients are decided.

For crack formation, if numerous non-interacting electrons in the structure go without interference in a sustained potential barrier, it can be presumed to be the equilibrium under the circumstance of constant entropy and volume (S, V) from Equation (3) in material. That is,

$$(dU)_{S,V} < 0 \qquad (8)$$

At that time, the Schrodinger's formulation in one dimension shall be stated:

$$-\frac{h^2}{8\pi^2 m}\frac{d^2\psi_n(x)}{dx^2} = E_n\psi_n \qquad (9)$$

where ψn is the wave function, E_n is the energy, h is the Planck constant, and m is the electron mass.

The boundary conditions are: 1) $x = 0$ or $x = a$ at barriers, $\psi n_{(}0)=\psi n(_a)=0$, 2) The chance of discovering the particle in a small room is $\int_0^a \psi^2(x)dx = 1$. Therefore, the solution can be quantized (David, 2018):

$$\psi_n(x) = \sqrt{\frac{2}{a}}\sin\left(\frac{n}{a}\right)x; \ E_n = \frac{n^2 h^2}{8ma^2} n > 0 \tag{10}$$

where $\psi n_{(}x+a)=\psi n(_x)$, a is the interval, and n is the principal quantum number.

As some effect is exerted, some transport processes happen in material. For instance, solid–state diffusion of impurities in silicon takes place: electromigration-caused voiding/growth of chloride ions and trapping of electrons or holes. That is, as an electromagnetic field (effect), ξ, is applied, solid-state diffusion of impurities in a silicon as semi-conductor material, $J1$ may be explained as follows (Grove,1967):

$$J_1 = \left[aC(x-a)\right]\cdot\exp\left[-\frac{q}{kT}\left(w-\frac{1}{2}a\xi\right)\right]\cdot v \tag{11}$$

where a is the interval between consecutive potential barriers, C is the concentration, q is the quantity of electric charge, w is potential barrier, ξ is the electromagnetic field, v is jump rate, $[aC(x-a)]$ is the density per unit area in the valley at $(x-a)$, and the exponential factor is the chance of a successful jump from the valley at $(x-a)$ to the valley at x.

A similar formula can be defined as $J_1, J_2,$ and J_3. As they are combined to express the flux J at location x, with the concentration $C(x\pm a)$ roughly by $C(x)\pm a(\partial C/\partial x)$, the flux J is

$$J = -\left[a^2 v e^{-qw/kT}\right]\cdot\cosh\frac{qa\xi}{2kT}\frac{\partial C}{\partial x} + \left[2ave^{-qw/kT}\right]C\sinh\frac{qa\xi}{2kT}$$
$$\cong \Phi(x,t,T)\sinh(a\xi)\exp\left(-\frac{Q}{kT}\right) \tag{12}$$

where k is the Boltzmann's constant, Q is the energy, and T is the temperature.

From statistical mechanics' standpoint, Equation (12) for given location, x, and temperature, T, in the steady state can be redefined as follows:

$$J \equiv B \sinh(a\xi) \exp\left(-\frac{Q}{kT}\right) \tag{13}$$

where Q is the energy, $\Phi()$ and B are constants.

Contrastingly, if $(dG)_{T,P}<0$ under the circumstance of constant temperature and pressure $(T = T_{surr}, P = P_{surr})$ from Equation (3), a chemical process that is controlled on speed can be formulated as:

$$K = K^+ - K^- = a\frac{kT}{h}e^{-\frac{\Delta E - \alpha S}{kT}} - a\frac{kT}{h}e^{-\frac{\Delta E + \alpha S}{kT}} = 2\frac{kT}{h}e^{-\frac{\Delta E}{kT}} \cdot \sinh\left(\frac{\alpha S}{kT}\right) = \sinh(aS)\exp\left(-\frac{\Delta E}{kT}\right) \tag{14}$$

If Equation (13) and (14) take reciprocal, the life-stress (LS) type can be defined:

$$TF = A[\sinh(aS)]^{-1} \exp\left(\frac{E_a}{kT}\right) \tag{15}$$

where A is constant.

As the hyperbolic expression considering stress may be changed into power or exponential function, Equation (15) is recommended as generalized prototype that can be explained such as fatigue, vibration, etc. That is, the term $[\sinh(aS)]^{-1}$ in Equation (15) can be expressed: (1). at low effect, $(S)^{-1}$ is almost linear, (2). at middle effect, $(S)^{-n}$, and (3). at high effect, $(e^{aS})^{-1}$. Now accelerated test scheme and approximate r-reliability and durability can be started.

An ALT is carried out in the region of medium stress (or effect). Because the power is defined as the multiplication of effort and flows, stresses in a multiport product can start from effort such as the induced vibration loads (F) (Karnopp, etc., 2012).

$$TF = A(S)^{-n} \exp\left(\frac{E_a}{kT}\right) \approx A'(e)^{-\lambda} \exp\left(\frac{E_a}{kT}\right) \approx B(F)^{-\lambda} \exp\left(\frac{E_a}{kT}\right) \tag{16}$$

where A' and B are constants, e is effort such as force, F.

The acceleration factor (AF) is stated as the relation between the standard state and the raised state as follows:

$$AF = \left(\frac{S_1}{S_0}\right)^n \left[\frac{E_a}{k}\left(\frac{1}{T_0} - \frac{1}{T_1}\right)\right] = \left(\frac{e_1}{e_0}\right)^\lambda \left[\frac{E_a}{k}\left(\frac{1}{T_0} - \frac{1}{T_1}\right)\right] = \left(\frac{F_1}{F_0}\right)^\lambda \left[\frac{E_a}{k}\left(\frac{1}{T_0} - \frac{1}{T_1}\right)\right] \tag{17}$$

where a_l is the elevated load level, a_o is the usual load level, T_o is the usual temperature, and T_l is the elevated temperature.

As raised tests are performed at the ambient temperature, Equation (17) is redefined as

$$AF = \left(\frac{S_1}{S_0}\right)^n = \left(\frac{e_1}{e_0}\right)^\lambda \tag{18}$$

Sample Size Formulation

To accomplish the desired assignment time of the ALT from the targeted BX life in the testing plan, the sample size formulation integrated with AF might be derived. The Weibull distribution for system life is extensively employed because it is defined as an expression of the characteristic life, η, and shape parameter, β. Therefore, if the system keeps to the Weibull distribution, the accumulative failure rate, $F(t)$, is defined:

$$F(t) = 1 - e^{-\left(\frac{t}{\eta}\right)^\beta} \tag{19}$$

where t is the (passed) time.

If Equation (19) takes the logarithm at $t=L_B$, it is

$$L_B^\beta = ln(1-x)^{-1} \eta_\alpha^\beta \tag{20}$$

where x is the accumulative failure rate until lifetime ($x=F/100$), and η_α^β is the characteristic life.

Failures on the Weibull distribution are divided into some classes—infant mortality, random failure, and wear-out failure—depending on shape parameter. The Weibayes procedure is explained as a Weibull examination with an assigned shape parameter, which can be obtained from prior experience or test data. Mentioning on Weibayes, characteristic life, $\eta M_{LE,}$ is attained from utilizing the maximum likelihood estimate (MLE):

$$\eta_{MLE}^\beta = \sum_{i=1}^{n} t_i^\beta / r \tag{21}$$

where t_i for each sample is testing time, and r is the failure numbers.

The confidence level is $100(1-a)$, so characteristic life, $\eta\alpha$ can be assessed as follows:

$$\eta_\alpha^\beta = \frac{2r}{\chi_\alpha^2(2r+2)} \cdot \eta_{MLE}^\beta = \frac{2}{\chi_\alpha^2(2r+2)} \cdot \sum_{i=1}^n t_i^\beta \tag{22}$$

If Equation (22) is substituted into Equation (20), it is

$$L_B^\beta = ln(1-x)^{-1} \frac{2}{\chi_\alpha^2(2r+2)} \cdot \sum_{i=1}^n t_i^\beta \tag{23}$$

As the whole reliability test is carried out with a limited sample number, the test scheme can be defined as:

$$nh^\beta \geq \sum t_i^\beta \geq (n-r)h^\beta \tag{24}$$

If Equation (24) is inserted into Equation (23), it is

$$L_B^\beta \geq ln(1-x)^{-1} \frac{2}{\chi_\alpha^2(2r+2)} \cdot (n-r)h^\beta \geq L_B^{*\beta} \tag{25}$$

If Equation (25) is reordered, the sample size expression is found as:

$$n \geq \frac{\chi_\alpha^2(2r+2)}{2} \times \frac{1}{ln(1-x)^{-1}} \times \left(\frac{L_B^*}{h}\right)^\beta + r \tag{26}$$

As the 1st term $\dfrac{\chi_\alpha^2(2r+2)}{2}$ in a 60% confidence level is approximated to $(r+1)$ and $ln\dfrac{1}{1-x}$ approximates to x, Equation (26) is redefined as:

$$n \geq (r+1) \times \frac{1}{x} \times \left(\frac{L_B^*}{h}\right)^\beta \tag{27}$$

As *AF* in Equation (18) is replaced into the test time, *h*, Equation (27) shall be redefined:

$$n \geq (r+1) \times \frac{1}{x} \times \left(\frac{L_B^*}{AF \cdot h_a} \right)^{\beta}$$ (28)

where Equation (31) will be clarified as *n* ~ (failed samples + 1)·(1/cumulative failure rate)·((objective life/(test time)) ^ *β*.

Equation (28) shall be affirmed as (Wasserman, 2003). Namely, for *n*≫*r*, the sample size shall be expressed as:

$$n = \frac{\chi_\alpha^2 (2r+2)}{2m^\beta lnR_L^{-1}} = \frac{\chi_\alpha^2 (2r+2)}{2} \times \frac{1}{ln(1-F_L)^{-1}} \times \left(\frac{L_B}{h} \right)^{\beta}$$ (29)

where $m \cong h/L_B$.

If *r*=0, the sample size can be indicated as:

$$n = \frac{ln(1-CL)}{m^\beta lnR_L} = \frac{-ln(1-CL)}{-m^\beta lnR_L} = \frac{ln(1-CL)^{-1}}{m^\beta lnR_L^{-1}} = \frac{ln\alpha^{-1}}{m^\beta lnR_L^{-1}} = \frac{\chi_\alpha^2 (2)}{2} \times \frac{1}{ln(1-F_L)^{-1}} \times \left(\frac{L_B}{h} \right)^{\beta}$$ (30)

where $2ln\alpha^{-1} = \chi_\alpha^2 (2)$ and *CL* is the confidence level.

If the objective of a product life–ice-maker is presumed to have a B1 life 10 years, the allocated test is computed for the assigned parts under raised circumstances. In executing parametric ALTs, the structural defects of a product operated by machinery will be found and altered to obtain the intended system life.

Case Study: Drawer in a French Door Refrigerator Exposed to Repeated Stress

Figure 3. French door refrigerator and freezer drawer assembly: (a) french door refrigerator; (b) mechanical components drawer: handle ①, drawer box ②, slide rail ③, and pocket box ④

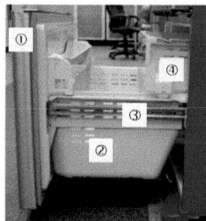

To keep the stored food fresh, a refrigerator supplies chilled air in its sections. As end-users hope to have decent entrance to the stocked food, A Freezer drawer in a French refrigerator is conceived to hold the indispensable food weights under predicted end-user circumstances in product life. Reserving food in the drawer therefore has the repeatedly handling ways: 1) opening the Freezer drawer to reserve food in the drawer, 2) grapping the stored food out of the Freezer drawer, and 3) closing it. Figure 3 shows a French door refrigerator with a newly designed drawer.

French-door refrigerators recalled from the marketplace because the handles of drawer cracked and fractured. As a result, end-users demanded the problematic refrigerators to be replaced because they no longer worked. Troublesome products indicated that the drawer had critical design imperfections in the construction. So, it might be redevised to endure repeated loading under end-user operation circumstances and enhance its reliability (Figure 4).

Figure 4. A damaged product after use

When watching the end-user use design of Freezer drawer, we knew that it would be repeatedly exposed to the food weight because of the working of the Freezer drawer. As the Freezer drawer had crucial design defects in the structure, designers had to replicate the problematic Freezer drawer experimentally and correct them.

Figure 5. Practical design idea of the freezer drawer: (a) drawer system, (b) free body diagram drawer

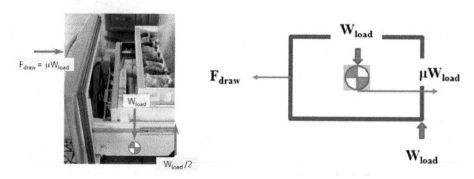

From the free-body diagram of the Freezer drawer in Figure 5, the force balance may be described as:

$$F_{draw} = \mu W l_{oad} \tag{31}$$

As the exerted stress in the Freezer drawer relies on the force proportionate to the stored food loads, the time-to-failure from Equation (16) might be described as:

$$TF = A(S)^{-n} = A(F_{draw})^{-\lambda} = A(\mu W_{load})^{-\lambda} \tag{32}$$

The acceleration factor (*AF*) in Equation (18) may be represented as:

$$AF = \left(\frac{S_1}{S_0}\right)^n = \left(\frac{F_1}{F_0}\right)^\lambda = \left(\frac{\mu W_1}{\mu W_0}\right)^\lambda = \left(\frac{W_1}{W_0}\right)^\lambda \tag{33}$$

For the Freezer drawer, the normal operating circumstances for an end-user compass from 0 to 43°C with humidity extending from 0% to 95%. In the US, the working of the drawer depends on end-user use. Market statics indicated that end-user operated the Freezer drawer in a refrigerator between five and nine cycles per day. With design lifetime for ten years, drawer occurred 36,500 use cycles.

For the worst occasion of the food load, the needed force was 0.34 kN (35 kg$_f$) on the handle of the drawer. The applied food load for the ALT was 0.68 kN (70 kg$_f$). Using Equation (33) with a quantity, λ, of 2, the total AF was approximately 4.0. For the targeted life of B1 life 10 years, the assigned cycles for three samples attained from Equation (28) were 67,000 cycles if the shape parameter, β, was supposed to have 2.0. The ALT was devised to assure a B1 life 10 years that it might fail less than once during 67,000 cycles. Figure 6 indicates the test arrangement of the ALT with labeled equipment for the design of Freezer drawer.

Figure 6. Accelerated life testing equipment: (a) ALT equipment, (b) controller

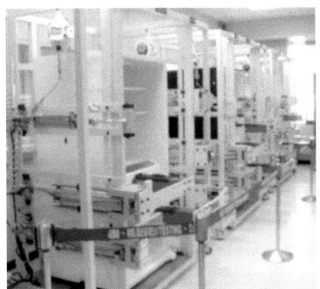 

RESULTS AND DISCUSSION

Refrigerators failed from the marketplace had dominant failure with cracked and fractured handle because of food weights in the operation of open/close of drawer. Market statics indicated that the problematic products would have design flaws. Due to these troublesome designs, the repetitive loads would cause unwanted stresses on the handle of drawer, leading it to crack, and finally generating the drawer failure (fracture). To replicate the failure of the Freezer drawer, ALTs were accomplished.

In the 1st ALT, the handle of the Freezer drawer cracked and fractured at 7,000 cycles and 8,000 cycles. Figure 7 indicates a photograph comparing the product failed from the field and the 1st ALT, separately. The shape of product failed the first ALT was indistinguishable to the ones returned from the marketplace. The test outcomes assured that the handle of the drawer was not well-devised to open and close the drawer door. So, by ALT, the troublesome handle of drawer should be replicated.

Figure 7. Failure of drawer handles in the field and 1st ALT outcome: (a) field, (b) 1st accelerated life testing results

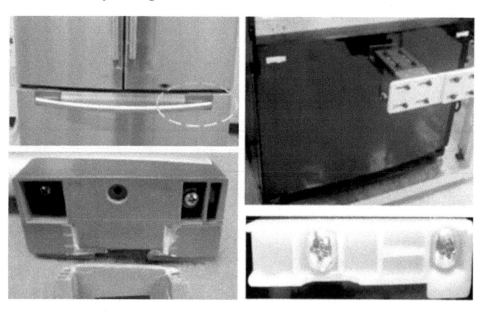

Figure 8. Altered handle of freezer drawer

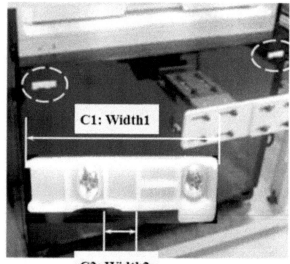

The fundamental causes of the cracked and fractured handle in the Freezer drawer came from its inadequate attached area. These design flaws could create the drawer handle to instantly be removed when exposed to repeated food weights. To cease the fractured handle from the Freezer drawer, the Freezer handle was changed: (1) enlarging the wideness of the strengthened handle, C1, from 90mm to 122mm; (2) enlarging the handle hooker size, C2, from 8mm to 19mm (Figure 8).

In the 2nd ALTs, the 10mm gap between the drawer and the Freezer compartment was found at 15,000 cycles and 16,000 cycles (Figure 9). As they were disassembled, component of slide rails in the Freezer drawer cracked and fractured. That is, as the food load for parametric ALT was repeatedly loaded, it was exposed that the slide rails were fragile because of their insufficient strength. These fractured rails came from the corner shape of the slide rails. As a result, the rail occurred to be cracked and fractured to its end. As action plans, the slide rail was modified: 1) enlarging the rail fastening screws, C3, from 1 to 2; 2) fastening the inner chamber and modifying material, C4, from HIPS to ABS; 3) thickening the boss, C5, from 2.0mm to 3.0mm; 4) fastening a new support rib, C6. By applying the ALT, the design of troublesome slide rail in the drawer was changed (Figure 9 & 10).

Figure 9. Cracked and fractured slide rails in the 2nd ALT: (a) fractured slide rail in freezer drawer, (b) detailed problem on the back of rail

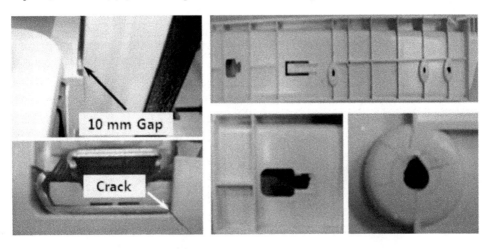

Figure 10. Redesigned slide rail

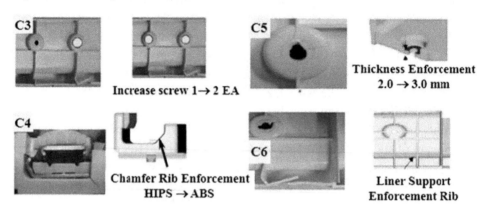

In the 3rd ALT, there were no matters until ALT was executed to 67,000 cycles. It thus was deduced that the alterations to the design discovered from the first ALT and second ALT were fulfilling a targeted aim. Table 2 abridges the ALT outcomes. With the regulated designs, the drawer is ensured to reach their lifetime aim – B1 life of ten years.

Table 2. Results of ALT

	1st ALT	2nd ALT	3rd ALT
	Initial Design	**Second Design**	**Final Design**
In 32,000 cycles, drawer has no crack	7,500 cycles: 2/6 Fracture 12,000 cycles: 1/3 OK	16,000 cycles: 2/3 Fracture	32,000 cycles: 3/3 OK 67,000 cycles: 3/3 OK
Structure (Drawer)	<handle> Fracture	<rail> 10 mm Gap Crack	
Material and specification	Width1: L90 →L122 Width2: L8→L19.0	Screw number: 1.0 → 2.0 Boss: 2.0 → 3.0 mm Chamfer1: Corner Materials: HIPS →ABS Adding a new support rib	

SUMMARY AND CONCLUSION

To recognize and alter the design issues of mechanical product such as the fatigue failure in a refrigerator, a structured reliability method based on new reliability notions, the fundamental vibration theory for developing quantum-transferred failure model, and sample size creating RQ specifications was supplied. It consisted of the following: 1) an ALT plan found on BX lifetime, 2) the vibrational load analysis for evolving the generalized life-stress model, 3) ALTs with modifications, and 4) judgement if the BX life objective is attained. As a case investigation, the reliability of a new Freezer drawer was studied.

- In the marketplace, the drawer in a refrigerator failed. As a close inspection, the fundamental cause of this failure in a refrigerator was determined to come from the improper design of handle. In the first ALT, the drawer handles exposed to repeated food weight cracked and fractured because of the design defect – inadequate fastened region of handle. As action plans, the drawer handle was redevised: (1) enlarging the witness of the strengthened handle, C1, from 90mm to 122mm; (2) enlarging the handle hooker size, C2, from 8mm to 19mm.

- During the second ALT, the fractured rails of Freezer drawer occurred because of their insufficient strength to withstand the repeated working of the drawer with the food weight. As action plans, the left/right rails in the drawer were redevised: 1) enlarging the rail fastening screw number, C3, from 1 to 2; 2) attaching an inner chamber and changing plastic material, C4, from HIPS to ABS; 3) thickening the boss, C5, from 2.0mm to 3.0mm; 4) attaching a new support rib, C6.

- In the third ALT, by modifying the designs of Freezer drawer. there were no difficulties. Thus, the newly devised Freezer drawers were anticipated to be the life need of B1 life 10 years.

Using this structured reliability methods with new notions, design issues such as fatigue of Freezer drawer in a new French door refrigerator can be resolved. This method may be pertinent to other mechanical products as well, including – cars, airplane, washing machines, construction machines, etc. If there are defects in the product where induced loads are exerted, the system shall not be successful during its predicted lifetime. By carrying out accelerated tests until the necessitated mission cycles (RQ specification), engineers can reproduce the failure and modify it.

REFERENCES

Anderson, T. L. (2017). *Fracture Mechanics—Fundamentals and Applications*. CRC. doi:10.1201/9781315370293

Bigg, G., & Billings, S. (2014). The iceberg risk in the *Titanic* year of 1912: Was it exceptional? *Significance*, *11*(3), 6–10. doi:10.1111/j.1740-9713.2014.00746.x

Branco, R., Prates, P., Costa, J. D. M., Berto, F., & Kotousov, A. (2018). New methodology of fatigue life evaluation for multiaxially loaded notched components based on two uniaxial strain-controlled tests. *International Journal of Fatigue*, *111*, 308–320. doi:10.1016/j.ijfatigue.2018.02.027

Campbell, F. C. (Ed.), *Elements of Metallurgy and Engineering Alloys*. doi:10.31399/asm.tb.emea.9781627082518

Cengel, Y., Boles, M., & Kanoglu, M. (2018). *Thermodynamics: An Engineering Approach* (9th ed.). McGraw—Hill.

David, J. G. (2018). *Introduction to Quantum Mechanics*. Cambridge University Press.

Duga, J. J., Fisher, W. H., Buxaum, R. W., Rosenfield, A. R., Buhr, A. R., Honton, E. J., & McMillan, S. C. (1982). *The Economic Effects of Fracture in the United States*. Final Report, September 30. Battelle Laboratories, Columbus, OH. Available as NBS Special Publication 647-2.

Elsayed, E. A. (2012). *Reliability Engineering*. John Wiley & Sons.

Garvin, D. A. (1987). Competing on the Eight Dimensions of Quality. *Harvard Business Review*, *65*(6), 101–109.

Goodno, B. J., & Gere, J. M. (2017). *Mechanics of Materials* (9th ed.). Cengage Learning, Inc.

Grove, A. (1967). *Physics and Technology of Semiconductor Device* (International Edition). Wiley.

Hahn, G. J., & Meeker, W. Q. (2004). *How to Plan an Accelerated Life Test (E-Book)*. ASQ Quality Press.

IEEE Standard Glossary of Software Engineering Terminology. (2002). *IEEE STD 610.12-1990. Standards Coordinating Committee of the Computer Society of IEEE*. Available online: https://ieeexplore.ieee.org/document/159342

Karnopp, D. C., Margolis, D. L., & Rosenberg, R. C. (2012). *System Dynamics: Modeling, Simulation, and Control of Mechatronic Systems*. John Wiley & Sons. doi:10.1002/9781118152812

Magaziner, I. C., & Patinkin, M. (1989). Cold competition: GE wages the refrigerator war. *Harvard Business Review*, *89*(2), 114–124.

McPherson, J. (1989) Accelerated Testing. In Electronic Materials Handbook; Volume 1: Packaging. ASM International Publishing.

Montgomery, D. (2013). *Design and Analysis of Experiments* (8th ed.). John Wiley and Son.

Palmgren, A. G. (1924). Die Lebensdauer von Kugellagern. *Z. Ver. Dtsch. Ing.*, *68*, 339–341.

Reddy, J. N. (2021). *An Introduction to Nonlinear Finite Element Method with Applications to Heat Transfer, Fluid Mechanics, and Solid Mechanics*. Oxford Press.

Taguchi, G., & Shih-Chung, T. (1992). *Introduction to quality engineering: bringing quality engineering upstream*. ASME.

Wasserman, G. (2003). *Reliability Verification, Testing, and Analysis in Engineering Design*. Marcel Dekker.

Woo, S., O'Neal, D., & Pecht, M. (2023). Improving the lifetime of mechanical systems during transit established on quantum/transport life-stress prototype and sample size. *Mechanical Systems and Signal Processing*, *193*, 110222. doi:10.1016/j.ymssp.2023.110222

Woo, S., & O'Neal, D. L. (2021). Reliability Design of Mechanical Systems Such as Compressor Subjected to Repetitive Stresses. *Metals*, *11*(8), 1261. doi:10.3390/met11081261

WRDA 2020 Updates. (2020). *The Final Report of the US House Committee on Transportation and Infrastructure on the Boeing 737 Max*. Available online: https://transportation.house.gov/committee-activity/boeing-737-max-investigation

Chapter 7

Navigating the Future of Ultra–Smart Computing Cyberspace:
Beyond Boundaries

N. Venkateswaran
Department of Master of Business Administration, Panimalar Engineering College, Chennai, India

Krishnamohan Reddy Kunduru
ⓘ https://orcid.org/0009-0009-8060-6216
Department of Engineering Design, Overhead Door Corporation, Lewisville, USA

Nanda Ashwin
Department of Computer Science and Engineering (IoT&CSBT), East Point College of Engineering and Technology, Bangalore, India

C. S. Sundar Ganesh
Department of Electrical and Electronics Engineering, Karpagam College of Engineering, Coimbatore, India

N. Hema
Department of Information Science and Engineering, RNS Institute of Technology, Bangalore, India

Sampath Boopathi
ⓘ https://orcid.org/0000-0002-2065-6539
Department of Mechanical Engineering, Muthayammal Engineering College (Autonomous), Namakkal, India

ABSTRACT

Ultra-smart computing cyberspace is a paradigm shift that combines artificial intelligence, augmented reality, and advanced networking technologies, transforming how we interact with digital environments. This integration offers unprecedented personalization, efficiency, and connectivity, blurring traditional computing boundaries and presenting challenges and opportunities in the ever-evolving technology landscape. Ultra-smart computing cyberspace presents opportunities

DOI: 10.4018/979-8-3693-2399-1.ch007

for creativity, collaboration, and commerce, but also presents challenges such as privacy concerns, cybersecurity threats, and ethical considerations. To address these, industry stakeholders, policymakers, and technologists must establish robust frameworks to safeguard user rights and ensure responsible innovation. However, by leveraging data-driven insights and human-centered design principles, organizations can unlock transformative value and stay ahead in the competitive digital landscape.

INTRODUCTION

Ultra-smart computing is a cutting-edge technology that combines AI, machine learning, IoT, quantum computing, and advanced data analytics to revolutionize data interaction and connectivity. It processes vast amounts of data at speeds unimaginable with traditional systems, enabling intelligent decisions, predictions, and recommendations. The integration of IoT devices further enhances the capabilities of ultra-smart computing by connecting physical objects to the digital realm, creating a network of interconnected devices that can communicate and share data in real-time. This fusion of technologies is driving unprecedented levels of automation, intelligence, and connectivity across various domains (Babulak, 2017a).

Ultra-smart computing is characterized by its ability to handle complexity and uncertainty with ease, unlike traditional systems that rely on predefined rules. It uses probabilistic models and deep learning techniques to extract meaningful insights from diverse data sources. The emergence of quantum computing adds another dimension to its capabilities, enabling calculations at exponentially faster speeds than classical computers. This quantum leap in processing power opens new possibilities for solving complex optimization problems, simulating molecular structures, and enhancing cryptography algorithms (Babulak, 2019).

Ultra-smart computing has applications in healthcare, finance, transportation, retail, and beyond. It enables personalized medicine through genetic analysis and predictive analytics, powers algorithmic trading strategies, and facilitates autonomous vehicles and traffic optimization in transportation. However, it also presents challenges like data privacy, algorithmic bias, cybersecurity threats, and the societal impact of automation, necessitating careful attention and proactive measures. Ultra-smart computing is crucial for organizations to stay competitive and innovate in the digital age (Babulak, 2021). Its ability to process and analyze vast amounts of data rapidly allows organizations to derive actionable insights, enhance decision-making processes, and create value. By leveraging artificial intelligence, machine learning, and advanced analytics, ultra-smart computing empowers businesses to stay agile, adapt to changing market dynamics, and capitalize on emerging opportunities. By responsibly and ethically embracing these transformative technologies, we can navigate the future of ultra-smart computing cyberspace and create a smarter, more connected, and sustainable world (Babulak, 2018a).

Ultra-smart computing is crucial for driving digital transformation, enabling organizations to modernize operations, streamline processes, and deliver superior customer experiences. It allows businesses to leverage data-driven strategies, optimize supply chain logistics, personalize marketing campaigns, and predict customer preferences. In the digital economy, data is the new currency, and the ability to harness and extract value from data assets is vital. Ultra-smart computing unlocks the full potential of data by uncovering hidden patterns, trends, and correlations, enabling strategic decision-making and driving business growth. It provides tools for anticipating market trends, mitigating risks, and capitalizing on opportunities proactively (Okamoto, 2021).

Ultra-smart computing has the potential to revolutionize various sectors, including healthcare, education, climate change, and urbanization, by utilizing AI and IoT technologies. It can improve healthcare outcomes, enhance educational experiences, and foster sustainable communities. The cyberspace landscape, which includes websites, social media networks, cloud computing infrastructure, and IoT devices, is a vast and dynamic ecosystem that shapes how we interact, transact, and engage with information online. This technology can drive meaningful social impact and improve healthcare outcomes, fostering a more sustainable and resilient society (Babulak, 2021; Okamoto, 2021).

The internet, at the core of the cyberspace landscape, is the foundation of digital connectivity and information exchange. With billions of interconnected devices, it's a vital tool for communication, commerce, and knowledge sharing, driving global economic growth and innovation. The cyberspace is characterized by various digital platforms and technologies, catering to different user needs. From social media platforms like Facebook to e-commerce giants like Amazon, the landscape provides numerous opportunities for businesses and individuals to connect and collaborate (Babulak, 2017a).

The rise of IoT devices and smart technologies has expanded the boundaries of the cyberspace landscape, encompassing smart homes, wearable devices, autonomous vehicles, and smart cities. This interconnected network of devices and sensors offers new opportunities for data collection, analysis, and automation, paving the way for more efficient and intelligent systems and services. By embracing the opportunities and challenges of ultra-smart computing, organizations and individuals can drive innovation, foster collaboration, and create a brighter, more interconnected future (Boopathi, 2024; Rahamathunnisa et al., 2024).

TRENDS SHAPING ULTRA-SMART COMPUTING

Artificial intelligence (AI) and machine learning (ML) are revolutionizing the future of ultra-smart computing. AI simulates human intelligence in machines, enabling them to perform tasks like learning, reasoning, and problem-solving. ML, a subset of AI, develops algorithms that allow computers to learn from data and improve over time without explicit programming. The integration of AI and ML in ultra-smart computing systems allows machines to analyze vast data, detect patterns, and make autonomous decisions with minimal human intervention (Babulak, 2017b).

AI and ML are revolutionizing healthcare, finance, and customer service by analyzing medical images, diagnosing diseases, and personalizing treatment plans. In finance, ML algorithms can predict stock prices, detect fraudulent transactions, and make informed investment decisions. In customer service, marketing, and product development, AI and ML are transforming customer engagement and driving growth through chatbots, virtual assistants, recommendation engines, and predictive analytics.

Internet of Things (IoT) Integration

The Internet of Things (IoT) is a significant trend in ultra-smart computing, enabling the integration of connected devices and sensors that collect and exchange data over the internet. This integration enables seamless communication between physical objects, allowing real-time interaction with users. IoT devices are transforming our interactions with the physical world, creating opportunities for automation, optimization, and innovation. Smart homes can remotely control IoT devices, enhancing convenience and energy efficiency (Babulak, 2018b).

IoT sensors in industrial and smart city systems can monitor performance, detect anomalies, and schedule maintenance, reducing downtime and optimizing operational efficiency. In smart cities, IoT technology improves traffic management, reduces energy consumption, and enhances public safety. Integrating IoT data with AI and ML algorithms allows organizations to gain insights, optimize processes, innovate products, and deliver personalized experiences. This data-driven approach drives intelligent decision-making and deeper understanding of user behavior (Boopathi, 2024; Hussain et al., 2023; Pachiappan et al., 2024).

The integration of AI, ML, and IoT is revolutionizing the field of ultra-smart computing, fostering innovation, efficiency, and connectivity across industries, thereby presenting new growth opportunities and value creation opportunities.

Quantum Computing Developments

Quantum computing is a revolutionary approach to computation that uses quantum mechanics principles to perform calculations at speeds far beyond those of classical computers. Despite being in its nascent stages, it holds immense potential to transform ultra-smart computing by addressing complex problems intractable for classical computers. One key development in quantum computing is the advancement in qubit technologies, which allow quantum computers to perform multiple calculations simultaneously, unlike classical bits which can only exist in a state of 0 or 1 (Rahamathunnisa et al., 2023).

Researchers are improving quantum error correction techniques to mitigate errors in quantum systems, which are susceptible to decoherence and environmental noise. They are also exploring new ways to use quantum computing for optimization, simulating quantum systems, and enhancing cryptography. Quantum computers have the potential to revolutionize drug discovery by simulating molecular interactions and predicting drug efficacy with unprecedented accuracy and speed. This progress is a significant development in quantum computing.

Quantum computing faces technical challenges like improving qubit coherence times, scaling qubits, and developing robust error correction schemes. Despite these obstacles, with continued investment and research from academia and industry, it is poised to revolutionize ultra-smart computing, unlocking new scientific, technological, and innovation frontiers.

Augmented Reality and Virtual Reality

Augmented and virtual reality (AR) and VR are immersive technologies that enhance human perception and interaction with digital content. They are transforming information, entertainment, and communication, and are expected to shape the future of ultra-smart computing. The adoption of AR and VR technologies has surged in various industries, including gaming, entertainment, education, healthcare, and retail. VR headsets and AR-enabled mobile apps blur the lines between the virtual and real worlds, providing users with unprecedented levels of immersion and engagement (Geng et al., 2022).

AR and VR have applications beyond entertainment, including training, simulation, remote collaboration, and experiential marketing. In healthcare, VR simulations improve medical skills and reduce errors. In retail, AR-powered apps allow customers to visualize products in real-world environments, enhancing shopping experiences and reducing returns. Advancements in hardware, such as lightweight headsets and high-resolution displays, make these technologies more accessible and user-friendly. As a result, AR and VR are increasingly integrated into everyday devices like smartphones, tablets, smart glasses, and wearable devices (Agrawal et al., 2023; Prabhuswamy et al., 2024; Sharma et al., 2024).

The integration of AR, VR, and AI and IoT technologies promises immersive and interactive experiences. From real-time contextual information to user-adaptive virtual reality training simulations, the future of AR and VR offers endless possibilities for innovation and creativity.

Cybersecurity Challenges

Implementing effective cybersecurity solutions can enhance organizations' resilience to cyber threats, safeguard assets, reputation, and customer trust in a digital world (Wilhelm et al., 2021).

- **Cyber Attacks**: The proliferation of sophisticated cyber threats, including malware, ransomware, phishing, and social engineering attacks, poses significant challenges to organizations worldwide. These attacks target vulnerabilities in networks, systems, and applications, leading to data breaches, financial losses, and reputational damage.
- **Data Privacy Concerns**: With the increasing volume of sensitive data stored and transmitted online, protecting data privacy has become a critical challenge. Compliance with regulations such as the General Data Protection Regulation (GDPR) and the California Consumer Privacy Act (CCPA) adds complexity to data management and requires organizations to implement robust security measures to safeguard personal information.
- **Insider Threats**: Insider threats, including malicious insiders and negligent employees, represent a significant cybersecurity risk for organizations. Insider threats can result in data exfiltration, sabotage, and intellectual property theft, posing serious challenges to maintaining the confidentiality, integrity, and availability of sensitive information.
- **Supply Chain Vulnerabilities**: The interconnected nature of modern supply chains introduces cybersecurity risks stemming from third-party vendors, suppliers, and partners. Supply chain attacks, such as supply chain compromise and software supply chain attacks, can exploit vulnerabilities in upstream and downstream dependencies, leading to widespread impact and disruption.
- **Emerging Technologies**: The adoption of emerging technologies, such as cloud computing, Internet of Things (IoT), and artificial intelligence (AI), introduces new cybersecurity challenges. Securing cloud environments, IoT devices, and AI-powered systems requires specialized expertise and proactive risk management strategies to mitigate potential threats and vulnerabilities.

Cybersecurity Solutions

- **Risk Management Frameworks**: Implementing robust risk management frameworks, such as the National Institute of Standards and Technology (NIST) Cybersecurity Framework or the ISO/IEC 27001 standard, helps organizations identify, assess, and mitigate cybersecurity risks effectively. These frameworks provide a structured approach to cybersecurity governance, risk assessment, and compliance management.
- **Multi-Layered Defense Strategies**: Adopting a multi-layered defense strategy that combines preventive, detective, and corrective controls helps organizations defend against a wide range of cyber threats. This approach may include endpoint protection, network segmentation, intrusion detection systems, and security information and event management (SIEM) solutions.
- **Employee Training and Awareness**: Investing in employee training and awareness programs helps build a security-aware culture within the organization and empowers employees to recognize and respond to cyber threats effectively. Training topics may include cybersecurity best practices, phishing awareness, password hygiene, and incident response procedures.
- **Encryption and Data Protection**: Implementing encryption and data protection measures, such as encryption at rest and in transit, helps safeguard sensitive information from unauthorized access and interception. Additionally, implementing data loss prevention (DLP) solutions enables organizations to monitor and control the movement of sensitive data across their networks.
- **Continuous Monitoring and Incident Response**: Establishing a robust cybersecurity monitoring and incident response program enables organizations to detect and respond to security incidents in a timely manner. Continuous monitoring of network traffic, system logs, and user activities helps identify anomalous behavior and potential security breaches, while incident response plans provide a structured approach to incident containment, eradication, and recovery (Malathi et al., 2024; Subha et al., 2023).

THE ROLE OF DATA IN ULTRA-SMART COMPUTING

Data lies at the heart of ultra-smart computing, serving as the fuel that powers AI algorithms, informs decision-making processes, and drives innovation across industries. In the context of ultra-smart computing, data is not just a byproduct of digital interactions but a strategic asset that organizations can leverage to gain insights, optimize operations, and create value in unprecedented ways (Babulak, 2018b).

Figure 1. Foundation data serves for ultra-smart computing

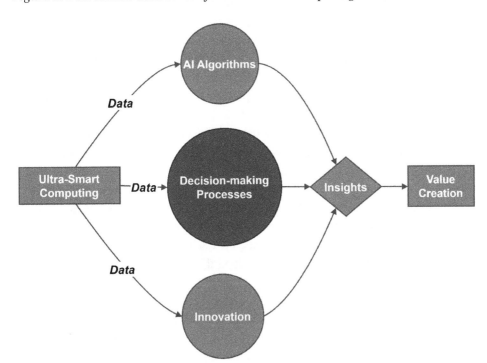

This Figure 1 illustrates how data serves as the foundation for ultra-smart computing, fueling AI algorithms, informing decision-making processes, and driving innovation, all of which ultimately lead to value creation.

Big Data Analytics

Big data analytics refers to the process of extracting actionable insights from large and complex datasets using advanced analytical techniques, such as machine learning, statistical analysis, and data mining. With the exponential growth of data generated from various sources, including sensors, devices, social media, and enterprise systems, big data analytics plays a crucial role in unlocking the value hidden within this vast sea of information (Gill et al., 2023).

- **Data Collection and Aggregation**: Big data analytics starts with the collection and aggregation of diverse data sources, including structured, unstructured, and semi-structured data. This may involve ingesting data from internal systems, external sources, and third-party data providers, as well as streaming data in real-time from IoT devices and sensors.

- **Data Processing and Transformation**: Once data is collected, it undergoes processing and transformation to prepare it for analysis. This may involve cleaning and filtering out irrelevant or duplicate data, harmonizing data from disparate sources, and transforming raw data into a format suitable for analysis.

- **Exploratory Data Analysis**: Exploratory data analysis (EDA) involves examining and visualizing the data to uncover patterns, trends, and correlations that may provide valuable insights. This step helps data analysts and data scientists gain a deeper understanding of the dataset and identify potential areas for further analysis.

- **Statistical Analysis and Modeling**: Big data analytics leverages statistical analysis and modeling techniques to identify relationships between variables, make predictions, and derive actionable insights from the data. This may involve running regression analysis, clustering algorithms, or classification models to uncover patterns and trends in the data.

- **Machine Learning and Predictive Analytics**: Machine learning algorithms play a central role in big data analytics, enabling organizations to build predictive models that can forecast future outcomes based on historical data. These models can be used for a wide range of applications, including customer segmentation, churn prediction, fraud detection, and demand forecasting (Maheswari et al., 2023).

- **Real-Time Analytics and Decision-Making**: In the era of ultra-smart computing, real-time analytics capabilities are essential for organizations to make data-driven decisions in the moment. Real-time analytics enable organizations to monitor key metrics, detect anomalies, and trigger automated responses in real-time, allowing them to respond quickly to changing market conditions and emerging threats (Revathi et al., 2024).

- **Continuous Improvement and Optimization**: Big data analytics is an iterative process that involves continuous improvement and optimization based on feedback and new data inputs. By leveraging insights gleaned from analytics, organizations can refine their strategies, optimize processes, and drive continuous innovation to stay ahead in a rapidly evolving digital landscape.

Big data analytics is crucial for organizations to utilize ultra-smart computing, unlock actionable insights, drive innovation, and create value from their data assets in today's data-driven world by utilizing advanced analytical techniques and technologies.

Data Privacy and Ethics

As organizations collect, analyze, and leverage vast amounts of data to fuel ultra-smart computing systems, ensuring data privacy and upholding ethical standards become paramount concerns. Data privacy regulations, such as the GDPR and CCPA, impose strict requirements on organizations to protect the privacy rights of individuals and safeguard their personal information. To address data privacy challenges in ultra-smart computing, organizations must implement robust data protection measures, such as encryption, access controls, and data anonymization, to mitigate the risk of unauthorized access and data breaches. Additionally, organizations must establish clear data governance policies and ethical guidelines to govern the collection, storage, and use of data in compliance with legal and ethical standards (Kannadhasan & Nagarajan, 2022).

Ethical considerations in ultra-smart computing extend beyond data privacy to encompass broader ethical implications, such as algorithmic bias, transparency, and accountability. Organizations must ensure that AI algorithms are trained on unbiased and representative data to avoid perpetuating discriminatory outcomes. Furthermore, transparency in AI decision-making processes and accountability for algorithmic decisions are essential to build trust and mitigate potential ethical concerns.

Data-driven Decision Making

Data-driven decision-making lies at the heart of ultra-smart computing, empowering organizations to derive actionable insights, optimize processes, and drive innovation based on data-driven insights. By leveraging AI and advanced analytics techniques, organizations can analyze large volumes of data in real-time to uncover patterns, trends, and correlations that inform strategic decision-making. In the realm of ultra-smart computing, data-driven decision-making enables organizations to personalize experiences, optimize operations, and anticipate market trends with greater precision and agility. For example, in the retail industry, data analytics can be used to segment customers based on their preferences and purchase history, allowing retailers to deliver targeted promotions and recommendations that drive sales and customer loyalty (Pitchai et al., 2024; Rebecca et al., 2024).

Data-driven decision-making helps organizations optimize resource allocation, mitigate risks, and capitalize on opportunities in a rapidly evolving business landscape. It uses predictive analytics and machine learning algorithms to forecast demand, optimize inventory, and identify supply chain disruptions. However, to fully utilize this technology, organizations must overcome challenges related to data quality, interoperability, and integration. Ensuring data accuracy, completeness, and consistency across diverse sources is crucial for informed decisions. Robust data integration and interoperability solutions are also necessary for seamless data flow and analysis.

In the era of ultra-smart computing, data privacy and ethics are crucial, necessitating robust data protection measures and ethical guidelines. Data-driven decision-making is vital for organizations to harness data's power, unlock innovation, efficiency, and competitive advantage in the digital age.

APPLICATIONS OF ULTRA-SMART COMPUTING

Figure 2. Various industries and their specific applications of ultra-smart computing

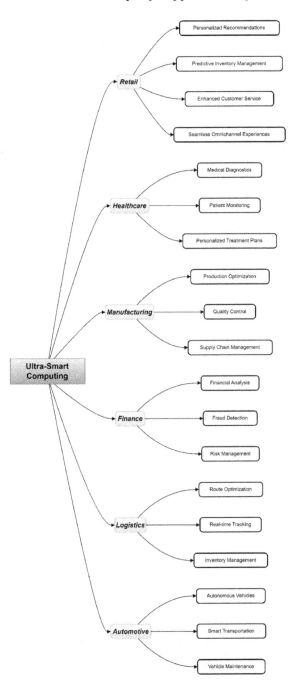

This Figure 2 illustrates various industries and their specific applications of ultra-smart computing, ranging from retail and healthcare to finance, logistics, manufacturing, and automotive sectors. Each industry is linked to specific applications or use cases enabled by ultra-smart computing technologies.

Smart Cities and Infrastructure

Ultra-smart computing is transforming urban environments into smart cities, transforming them into efficient, sustainable ecosystems. It uses advanced technologies like IoT, AI, and data analytics to optimize resource allocation, enhance public services, and improve residents' quality of life. IoT sensors collect real-time data on traffic flow, air quality, energy consumption, and waste management, which are then analyzed using AI algorithms for actionable insights and decision-making to improve urban planning, transportation systems, and environmental sustainability (Maguluri et al., 2023; Sonia et al., 2024).

Ultra-smart computing is transforming infrastructure solutions, enhancing resilience, reliability, and efficiency in critical systems like energy grids, water distribution networks, and public transportation. By integrating IoT sensors, predictive analytics, and automation technologies, cities can monitor assets, detect potential failures, and optimize resource usage. Smart cities also use ultra-smart computing to improve public safety and security through advanced surveillance systems, emergency response mechanisms, and predictive policing algorithms. By analyzing data from various sources, cities can identify security threats, respond effectively, and prevent crime (Maguluri et al., 2023; Sonia et al., 2024).

Healthcare Innovations

Ultra-smart computing is revolutionizing the healthcare industry by providing innovative solutions that improve patient care, enhance clinical outcomes, and streamline delivery processes. It enables personalized medicine, remote patient monitoring, predictive analytics, and medical imaging, allowing healthcare providers to analyze genomic data, medical records, and other patient-specific information to tailor treatment plans and interventions. AI algorithms and predictive analytics help identify patterns in patient data, predict disease risk, optimize treatment efficacy, and improve patient outcomes (Boopathi & Khang, 2023; Kushwah et al., 2024; Malathi et al., 2024).

Ultra-smart computing allows healthcare providers to monitor patients' vital signs, medication adherence, and health status remotely. By connecting wearable devices, IoT sensors, and mobile apps, healthcare professionals can detect early signs of deteriorating health and intervene proactively to prevent hospital readmissions. It also powers medical imaging technologies like MRI, CT scans, and X-rays, enabling high-resolution images for diagnosis and treatment planning. AI algorithms for image analysis and interpretation help radiologists detect abnormalities, classify diseases, and prioritize cases, reducing diagnosis times and improving patient care (Ali et al., 2024; Sonia et al., 2024).

Ultra-smart computing is being utilized in smart cities, infrastructure, and healthcare innovations to address complex challenges and drive positive societal impact. This technology can enhance efficiency, sustainability, and livability, while healthcare organizations can provide personalized, proactive care, improving health outcomes and quality of life.

Financial Services and Fintech

Ultra-smart computing applications in financial services and fintech improve trading strategies, risk management, fraud detection, and personalized services. In transportation and logistics, it optimizes route planning, fleet management, supply chain operations, and autonomous vehicle development, leading to cost savings, operational efficiency improvements, and increased customer satisfaction (Rahamathunnisa et al., 2024; Ravisankar et al., 2023).

- **Algorithmic Trading**: Ultra-smart computing enables financial institutions to leverage AI and machine learning algorithms to analyze market data, identify trading opportunities, and execute trades at high speeds. Algorithmic trading systems can process vast amounts of data in real-time, enabling automated trading strategies to capitalize on market inefficiencies and fluctuations.
- **Risk Management**: Ultra-smart computing enhances risk management capabilities in the financial services industry by enabling organizations to analyze and assess risk factors more accurately and dynamically. AI-powered risk models can analyze historical data, market trends, and external factors to identify potential risks and predict their impact on portfolios, enabling organizations to make informed decisions and mitigate risk exposure.
- **Fraud Detection**: Ultra-smart computing plays a crucial role in fraud detection and prevention within the financial services sector. Machine learning algorithms can analyze transactional data, user behavior patterns, and anomalies to identify suspicious activities and potential fraudulent transactions in real-time. By leveraging AI-powered fraud detection systems, financial institutions can detect and prevent fraud more effectively while minimizing false positives and preserving customer trust.
- **Personalized Financial Services**: Ultra-smart computing enables financial institutions to deliver personalized financial services tailored to individual

customer preferences, needs, and behaviors. AI algorithms can analyze customer data, including transaction history, demographics, and financial goals, to offer personalized recommendations for banking products, investment opportunities, and financial planning services. Personalization enhances customer engagement, loyalty, and satisfaction, driving business growth and competitive advantage.

Transportation and Logistics Optimization

- **Route Optimization**: Ultra-smart computing optimizes transportation and logistics operations by analyzing vast amounts of data, including traffic patterns, weather conditions, and delivery schedules, to optimize route planning and vehicle dispatching. AI-powered route optimization algorithms can identify the most efficient routes, minimize delivery times, and reduce fuel consumption, leading to cost savings and improved operational efficiency.
- **Fleet Management**: Ultra-smart computing enables organizations to optimize fleet management operations by leveraging IoT sensors, telematics, and predictive analytics. IoT-enabled devices installed in vehicles collect real-time data on vehicle performance, fuel consumption, and driver behavior, enabling organizations to monitor fleet health, schedule maintenance proactively, and optimize vehicle utilization. Predictive analytics algorithms can forecast maintenance needs, predict equipment failures, and optimize fleet routing to minimize downtime and enhance productivity.
- **Supply Chain Optimization**: Ultra-smart computing enhances supply chain management by optimizing inventory management, demand forecasting, and logistics operations. AI algorithms analyze historical sales data, market trends, and external factors to forecast demand accurately, enabling organizations to optimize inventory levels, reduce stockouts, and minimize excess inventory holding costs. Furthermore, AI-powered predictive analytics can optimize supply chain logistics by identifying bottlenecks, optimizing transportation routes, and improving warehouse operations, leading to cost savings and operational efficiency improvements (Mohanty, Venkateswaran, et al., 2023; Verma et al., 2024).
- **Autonomous Vehicles**: Ultra-smart computing enables the development and deployment of autonomous vehicles, including self-driving cars, trucks, drones, and delivery robots. AI algorithms process sensor data from cameras, lidar, radar, and GPS to perceive the surrounding environment, make real-time decisions, and navigate autonomously. Autonomous vehicles have the potential to revolutionize transportation and logistics by improving safety,

reducing congestion, and enhancing efficiency in goods and passenger transportation (Dhanalakshmi et al., 2024).

Retail and Customer Experience Enhancement

Ultra-smart computing is revolutionizing the retail industry by enhancing customer experience, personalizing interactions, and driving sales through data-driven strategies and innovative technologies, reshaping customer engagement and enabling retailers to differentiate themselves in a competitive market (Revathi et al., 2024).

- Personalized Recommendations: Ultra-smart computing allows retailers to analyze customer data, such as purchase history and browsing habits, to create personalized product recommendations. This, using machine learning algorithms and predictive analytics, increases conversion rates and customer satisfaction. These recommendations can be delivered across various touchpoints, ensuring a seamless shopping experience for customers.
- Predictive Inventory Management: Ultra-smart computing allows retailers to optimize inventory levels and anticipate demand fluctuations by analyzing historical sales data, market trends, and external factors. AI-driven demand forecasting and inventory optimization algorithms minimize stockouts, reduce excess inventory, and improve supply chain efficiency. This predictive inventory management enhances product availability and reduces costs associated with overstocking and understocking.
- Enhanced Customer Service: Ultra-smart computing is enabling the development of intelligent virtual assistants, chatbots, and voice-enabled devices that improve customer service and support. These tools, using natural language processing and sentiment analysis algorithms, automate customer inquiries, offer personalized assistance, and resolve issues in real-time, freeing human agents to focus on more complex customer needs (Revathi et al., 2024).
- Seamless Omnichannel Experiences: Ultra-smart computing allows retailers to provide omnichannel experiences, integrating data from various sources like e-commerce platforms, mobile apps, social media, and physical stores. This ensures consistent, personalized experiences throughout the customer journey. It also facilitates features like click-and-collect, BOPIS, and ship-from-store, allowing retailers to offer flexible fulfillment options and meet customer expectations for convenience and flexibility.

Ultra-smart computing is revolutionizing the retail industry by improving customer experience, optimizing operations, and driving business growth. It offers personalized recommendations, predictive inventory management, enhanced customer service, and seamless omnichannel experiences, enabling retailers to stand out in a competitive market and meet consumer needs.

CHALLENGES AND OPPORTUNITIES IN ULTRA-SMART COMPUTING

Ultra-smart computing presents ethical and societal challenges, including regulatory compliance, talent acquisition, and infrastructure development. However, these challenges also present opportunities for organizations to demonstrate leadership, foster innovation, and drive positive societal impact. By proactively addressing these challenges, organizations can unlock its full potential for value creation and sustainable growth in the digital age (Prabhuswamy et al., 2024; Puranik et al., 2024; Venkatasubramanian et al., 2024). This Figure 3 illustrates the continuous cycle between challenges and opportunities in ultra-smart computing. It depicts the challenges such as ethical concerns, data privacy, security risks, and compliance issues, which lead to opportunities for innovation, competitive advantage, business growth, and market leadership. These opportunities, in turn, present new challenges, forming a continuous cycle of adaptation and improvement in ultra-smart computing.

- Ethical and Societal Implications: The rapid advancement of ultra-smart computing raises ethical concerns about data privacy, algorithmic bias, and automation's impact on employment and society. Addressing these issues presents opportunities for organizations to demonstrate responsible AI and data governance practices, build stakeholder trust, and mitigate reputational risks (Boopathi & Khang, 2023).
- Regulatory and Legal Considerations: The regulatory landscape for ultra-smart computing is complex, involving data protection laws, industry standards, and AI regulations. Addressing these considerations proactively can ensure compliance, minimize risks, and promote transparency and accountability in organizations using these technologies.
- Talent and Skill Requirements: The rapid growth of ultra-smart computing necessitates organizations to recruit and retain skilled professionals in data science, machine learning, cybersecurity, and AI ethics. However, investing in talent development and upskilling can create a workforce capable of utilizing this technology, driving innovation, and maintaining a competitive edge in the digital economy.
- Infrastructure and Connectivity Issues: Ultra-smart computing faces challenges in regions with limited broadband internet access due to its reliance

on robust infrastructure and high-speed connectivity. However, investing in infrastructure development and expanding connectivity initiatives like 5G networks can enhance accessibility, unlock new markets, and promote economic growth.

Figure 3. Continuous cycle between challenges and opportunities in ultra-smart computing

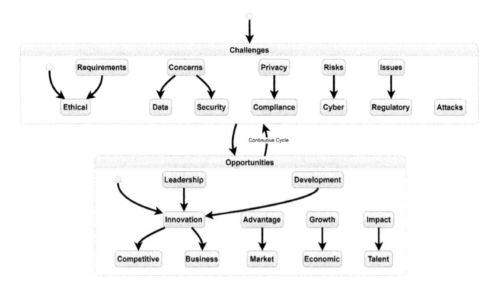

STRATEGIES FOR SUCCESS IN ULTRA-SMART COMPUTING

Organizations can capitalize on the full potential of ultra-smart computing by adopting strategies such as collaborative ecosystem development, investment in research and development, continuous learning, and building resilient systems. These strategies enable them to navigate the complexities of this rapidly evolving landscape and drive innovation (Sangeetha et al., 2023; Venkateswaran, Vidhya, et al., 2023).

Figure 4. Important strategies for ultra-smart computing success

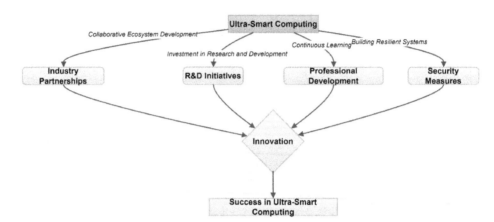

Figure 4 outlines important strategies for ultra-smart computing success, including collaborative ecosystem development, investment in research, continuous learning, and resilient systems, which drive innovation and ultimately lead to success in the technology industry.

- Collaborative Ecosystem Development: The text emphasizes the importance of fostering partnerships with industry peers, academic institutions, research organizations, and technology providers to foster innovation and knowledge sharing. Strategic alliances can address common challenges and accelerate the development of ultra-smart computing technologies. Engaging with startups and accelerators can identify emerging technologies and innovative solutions.

- Investment in Research and Development: The initiative allocates resources and investment towards research and development initiatives focusing on ultra-smart computing technologies like AI, machine learning, quantum computing, and IoT. It establishes dedicated teams and innovation labs to explore new ideas and collaborates with academic institutions to fund joint projects, sponsor scholarships, and recruit top talent.

- Continuous Learning and Adaptation: To foster a culture of continuous learning and adaptation in your organization, encourage employees to stay updated on ultra-smart computing trends and technologies. Provide training and professional development opportunities to equip employees with the necessary skills. Encourage experimentation and risk-taking, empowering teams to explore new ideas, iterate solutions, and learn from failures for innovation and excellence.

- Building Resilient and Secure Systems: The text emphasizes the importance of prioritizing cybersecurity and data privacy in the design and implementation of ultra-smart computing systems, integrating security controls and best practices throughout the development lifecycle. It also highlights the need for robust security measures like encryption, access controls, and intrusion detection systems to protect sensitive data and mitigate cyber threats.
- Continuous Learning and Adaptation: To stay competitive in the ultra-smart computing landscape, companies should adopt a culture of continuous learning and adaptation. This involves investing in training programs in areas like artificial intelligence, machine learning, cybersecurity, and data governance. Fostering a culture of innovation encourages employees to explore new ideas and technologies. Staying informed about emerging technologies and market dynamics through industry conferences and networking events helps companies adapt strategies and tactics to maintain a competitive edge (Dhanalakshmi et al., 2024; Maheswari et al., 2023).
- Building Resilient and Secure Systems: The text emphasizes the importance of prioritizing cybersecurity and data protection initiatives in ultra-smart computing systems to protect sensitive information, mitigate risks, and comply with regulatory requirements. A multi-layered security approach, including preventive, detective, and corrective controls, is recommended. Regular security assessments, penetration testing, and vulnerability scans are also recommended. Collaboration among internal teams, external partners, and industry stakeholders is encouraged to enhance threat intelligence capabilities. A proactive approach to incident response and crisis management is adopted, focusing on developing plans, playbooks, and communication protocols to minimize the impact of security incidents and ensure business continuity in case of cyber or data breaches (Boopathi & Khang, 2023).

By prioritizing continuous learning and adaptation and building resilient and secure systems, organizations can position themselves for success in the rapidly evolving landscape of ultra-smart computing. These strategies enable organizations to stay agile, innovative, and competitive while effectively managing risks and ensuring the integrity and security of their ultra-smart computing initiatives.

BEST PRACTICES

Implementing ultra-smart computing solutions across industries can drive business value, unlock new opportunities for innovation, efficiency, and growth in the digital age by following best practices (Babulak, 2018a; Okamoto, 2021).

- **Start with Clear Business Objectives**: Before implementing ultra-smart computing solutions, it's essential to define clear business objectives and align them with strategic goals. Identify specific use cases and areas where ultra-smart computing can drive value, whether it's improving operational efficiency, enhancing customer experiences, or optimizing decision-making processes.

- **Data Quality and Governance**: Ensure that the data used for ultra-smart computing initiatives is of high quality, accurate, and relevant to the intended objectives. Establish robust data governance practices to maintain data integrity, security, and compliance with regulatory requirements. Implement data quality checks, data validation processes, and data cleansing techniques to ensure that data inputs are reliable and consistent.

- **Cross-Functional Collaboration**: Foster collaboration among cross-functional teams, including data scientists, software engineers, business analysts, and domain experts, to leverage diverse perspectives and expertise in designing and implementing ultra-smart computing solutions. Encourage open communication and knowledge sharing to facilitate innovation and problem-solving across organizational silos (Puranik et al., 2024; Revathi et al., 2024).

- **Agile Development Methodologies**: Adopt agile development methodologies, such as Scrum or Kanban, to iteratively design, develop, and deploy ultra-smart computing solutions. Break down complex projects into smaller, manageable tasks, and prioritize features based on business value and stakeholder feedback. Embrace an iterative approach to development, allowing for rapid prototyping, testing, and iteration to adapt to changing requirements and user needs.

- **User-Centric Design**: Prioritize user experience (UX) and design principles in the development of ultra-smart computing solutions to ensure that they are intuitive, user-friendly, and aligned with user expectations and preferences. Conduct user research, usability testing, and feedback sessions to gather insights and iterate on the design based on user feedback and behavior.

- **Scalability and Flexibility**: Design ultra-smart computing solutions with scalability and flexibility in mind to accommodate future growth and evolving business requirements. Choose scalable architectures, cloud-based platforms, and modular design patterns that can easily adapt to changing data volumes, user demands, and technological advancements.

- **Robust Security Measures**: Implement robust security measures to protect ultra-smart computing systems and data assets from cyber threats, breaches, and unauthorized access. Apply encryption, access controls, and authentication mechanisms to secure sensitive data and prevent data breaches. Regularly update security patches and conduct security audits and vulnerability assessments to identify and mitigate security risks proactively.

- **Performance Monitoring and Optimization**: Establish performance monitoring and optimization processes to ensure that ultra-smart computing solutions operate efficiently and effectively. Monitor key performance indicators (KPIs), such as processing speed, accuracy, and resource utilization, and identify areas for improvement through performance tuning, optimization algorithms, and infrastructure scaling (Mohanty, Jothi, et al., 2023; Pramila et al., 2023).

- **Continuous Improvement and Innovation**: Foster a culture of continuous improvement and innovation by encouraging experimentation, learning from failures, and embracing new technologies and methodologies. Invest in research and development initiatives, collaborate with industry partners, and participate in innovation ecosystems to stay ahead of the curve and drive continuous innovation in ultra-smart computing.

- **Compliance and Ethical Considerations**: Ensure compliance with regulatory requirements, industry standards, and ethical guidelines governing the use of ultra-smart computing technologies. Adhere to data privacy regulations, such as GDPR and CCPA, and uphold ethical principles, such as fairness, transparency, and accountability, in the collection, processing, and use of data for ultra-smart computing initiatives.

FUTURE OUTLOOK AND PREDICTIONS

Ultra-smart computing is poised for significant transformations with the emergence of quantum computing, edge computing, and synthetic biology. These technologies can significantly impact society and business. By responsibly embracing these technologies, addressing ethical concerns, and collaborating with stakeholders, organizations can harness their power for a sustainable and inclusive future (Babulak, 2018a, 2018b; Kannadhasan & Nagarajan, 2022).

Figure 5. Ultra-smart computing, highlighting the transformative impact of emerging technologies

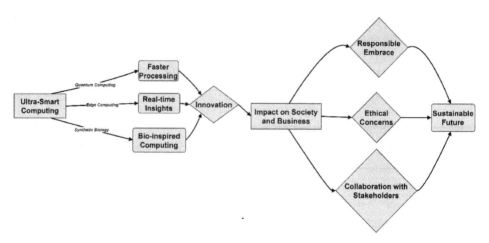

Figure 5 presents the future outlook for ultra-smart computing, highlighting the transformative impact of emerging technologies like quantum computing, edge computing, and synthetic biology. It emphasizes the need for responsible embracing, ethical concerns, and stakeholder collaboration for a sustainable future.

Emerging Technologies on the Horizon

- Quantum Computing: Quantum computing is poised to revolutionize ultra-smart computing by enabling exponential gains in processing power and solving complex problems that are currently intractable for classical computers. As quantum computing technologies mature, they will unlock new possibilities for simulation, optimization, and cryptography, driving innovation across industries.

- Edge Computing: Edge computing, which involves processing data closer to the source or endpoint devices, is gaining momentum as organizations seek to leverage real-time insights and reduce latency in data processing. By distributing computing resources across the network edge, edge computing enables faster decision-making, enhanced security, and improved scalability for ultra-smart computing applications (Maguluri et al., 2023).

- Synthetic Biology: Synthetic biology, the engineering of biological systems for practical purposes, holds promise for ultra-smart computing through the development of bio-inspired computing systems and biocomputing architectures. By harnessing the power of biological processes, such as DNA computing and neural networks, synthetic biology could lead to breakthroughs in energy efficiency, data storage, and computational capabilities (Reddy et al., 2023; Venkateswaran, Kumar, et al., 2023).

Potential Impact on Society and Business

- Economic Disruption: The widespread adoption of ultra-smart computing technologies is expected to drive economic disruption by transforming industries, reshaping business models, and creating new opportunities for innovation and entrepreneurship. However, this disruption may also lead to job displacement and workforce transitions as automation and AI-driven technologies replace traditional roles and tasks.
- Societal Transformation: Ultra-smart computing has the potential to transform society by improving healthcare outcomes, enhancing education, and addressing societal challenges such as climate change and poverty. By leveraging AI, IoT, and advanced analytics, ultra-smart computing can enable more personalized healthcare services, adaptive learning experiences, and sustainable solutions for environmental conservation.
- Ethical Considerations: As ultra-smart computing technologies become more pervasive, ethical considerations surrounding data privacy, algorithmic bias, and societal impact will become increasingly important. Organizations must prioritize ethical principles, such as fairness, transparency, and accountability, in the development and deployment of ultra-smart computing solutions to ensure that they benefit society while minimizing potential risks and harms (Boopathi & Khang, 2023).
- Regulatory Frameworks: Governments and regulatory bodies are likely to introduce new regulations and policies to govern the use of ultra-smart computing technologies and address emerging risks and challenges. Regulations may focus on data privacy, cybersecurity, AI ethics, and responsible innovation to ensure that ultra-smart computing advances societal goals while upholding fundamental rights and values.

CONCLUSION

The chapter on ultra-smart computing offers a comprehensive overview of the potential, challenges, and opportunities of using advanced technologies to drive innovation and value creation in the digital age, exploring key concepts, applications, best practices, and future outlook.

- Ultra-smart computing represents a paradigm shift in how organizations harness data, intelligence, and connectivity to achieve strategic objectives, enhance competitiveness, and create value across industries.

- The convergence of technologies such as artificial intelligence, machine learning, IoT, and advanced analytics is driving unprecedented innovation and disruption, reshaping business models, and redefining the way we live, work, and interact in society.

- While ultra-smart computing offers immense potential for driving economic growth, improving quality of life, and addressing societal challenges, it also presents complex ethical, legal, and societal implications that must be addressed proactively and responsibly.

- Successful implementation of ultra-smart computing requires organizations to adopt a holistic approach that encompasses clear business objectives, robust data governance, cross-functional collaboration, agile methodologies, user-centric design, and a commitment to security, compliance, and ethical principles.

- Looking ahead, emerging technologies such as quantum computing, edge computing, and synthetic biology hold promise for unlocking new frontiers in ultra-smart computing, driving further innovation, and transforming society and business in profound ways.

- As organizations navigate the evolving landscape of ultra-smart computing, it is essential to prioritize continuous learning, adaptation, and innovation, foster a culture of collaboration and transparency, and embrace responsible practices that balance technological advancement with ethical considerations and societal impact.

The chapter emphasizes the transformative power of ultra-smart computing in driving innovation, enhancing competitiveness, and creating positive societal impact. It emphasizes the need for responsible stewardship, ethical governance, and strategic foresight to fully realize its potential in the digital age, enabling organizations to thrive in a data-driven world.

REFERENCES

Agrawal, A. V., Pitchai, R., Senthamaraikannan, C., Balaji, N. A., Sajithra, S., & Boopathi, S. (2023). Digital Education System During the COVID-19 Pandemic. In Using Assistive Technology for Inclusive Learning in K-12 Classrooms (pp. 104–126). IGI Global. doi:10.4018/978-1-6684-6424-3.ch005

Ali, M. N., Senthil, T., Ilakkiya, T., Hasan, D. S., Ganapathy, N. B. S., & Boopathi, S. (2024). IoT's Role in Smart Manufacturing Transformation for Enhanced Household Product Quality. In *Advanced Applications in Osmotic Computing* (pp. 252–289). IGI Global. doi:10.4018/979-8-3693-1694-8.ch014

Babulak, E. (2017). Cyber Security Solutions and Challenges in Ultrafast Internet Connecting Ultra-Smart Computational Devices. *ICSES Transactions on Computer Networks and Communications, 3*(1).

BabulakE. (2018). Role of Computational Physics in the Age Third Millennium Automation. *Available at* SSRN 3404595.

Babulak, E. (2019). Third Millennium Smart Cyberspace. *ICSES, 3,* 1–2.

Babulak, E. (2021). Third Millennium Life Saving Smart Cyberspace Driven by AI & Robotics. *COJ Rob Artificial Intel, 1*(4).

Boopathi, S. (2024). Sustainable Development Using IoT and AI Techniques for Water Utilization in Agriculture. In Sustainable Development in AI, Blockchain, and E-Governance Applications (pp. 204–228). IGI Global. doi:10.4018/979-8-3693-1722-8.ch012

Boopathi, S., & Khang, A. (2023). AI-Integrated Technology for a Secure and Ethical Healthcare Ecosystem. In *AI and IoT-Based Technologies for Precision Medicine* (pp. 36–59). IGI Global. doi:10.4018/979-8-3693-0876-9.ch003

Dhanalakshmi, M., Tamilarasi, K., Saravanan, S., Sujatha, G., Boopathi, S., & Associates. (2024). Fog Computing-Based Framework and Solutions for Intelligent Systems: Enabling Autonomy in Vehicles. In Computational Intelligence for Green Cloud Computing and Digital Waste Management (pp. 330–356). IGI Global.

Geng, R., Li, M., Hu, Z., Han, Z., & Zheng, R. (2022). Digital Twin in smart manufacturing: Remote control and virtual machining using VR and AR technologies. *Structural and Multidisciplinary Optimization, 65*(11), 321. doi:10.1007/s00158-022-03426-3

Gill, A., Khalid, A., & Murugan, R. (2023). A Crucial Review on Ai Use In Hybrid Grids: A Ultra Smart Way Towards Micro Grid. *2023 International Conference on Power Energy, Environment & Intelligent Control (PEEIC)*, 1364–1368. 10.1109/PEEIC59336.2023.10451299

Hussain, Z., Babe, M., Saravanan, S., Srimathy, G., Roopa, H., & Boopathi, S. (2023). Optimizing Biomass-to-Biofuel Conversion: IoT and AI Integration for Enhanced Efficiency and Sustainability. In Circular Economy Implementation for Sustainability in the Built Environment (pp. 191–214). IGI Global.

Kannadhasan, S., & Nagarajan, R. (2022). Recent trends in wearable technologies, challenges and opportunities. *Designing Intelligent Healthcare Systems, Products, and Services Using Disruptive Technologies and Health Informatics*, 131–143.

Kushwah, J. S., Gupta, M., Shrivastava, S., Saxena, N., Saini, R., & Boopathi, S. (2024). Psychological Impacts, Prevention Strategies, and Intervention Approaches Across Age Groups: Unmasking Cyberbullying. In Change Dynamics in Healthcare, Technological Innovations, and Complex Scenarios (pp. 89–109). IGI Global.

Maguluri, L. P., Arularasan, A., & Boopathi, S. (2023). Assessing Security Concerns for AI-Based Drones in Smart Cities. In Effective AI, Blockchain, and E-Governance Applications for Knowledge Discovery and Management (pp. 27–47). IGI Global. doi:10.4018/978-1-6684-9151-5.ch002

Maheswari, B. U., Imambi, S. S., Hasan, D., Meenakshi, S., Pratheep, V., & Boopathi, S. (2023). Internet of things and machine learning-integrated smart robotics. In Global Perspectives on Robotics and Autonomous Systems: Development and Applications (pp. 240–258). IGI Global. doi:10.4018/978-1-6684-7791-5.ch010

Malathi, J., Kusha, K., Isaac, S., Ramesh, A., Rajendiran, M., & Boopathi, S. (2024). IoT-Enabled Remote Patient Monitoring for Chronic Disease Management and Cost Savings: Transforming Healthcare. In Advances in Explainable AI Applications for Smart Cities (pp. 371–388). IGI Global.

Mohanty, A., Jothi, B., Jeyasudha, J., Ranjit, P., Isaac, J. S., & Boopathi, S. (2023). Additive Manufacturing Using Robotic Programming. In *AI-Enabled Social Robotics in Human Care Services* (pp. 259–282). IGI Global. doi:10.4018/978-1-6684-8171-4.ch010

Mohanty, A., Venkateswaran, N., Ranjit, P., Tripathi, M. A., & Boopathi, S. (2023). Innovative Strategy for Profitable Automobile Industries: Working Capital Management. In Handbook of Research on Designing Sustainable Supply Chains to Achieve a Circular Economy (pp. 412–428). IGI Global.

Okamoto, E. (2021). Overview of nonlinear signal processing in 5G and 6G access technologies. *Nonlinear Theory and Its Applications, IEICE*, 12(3), 257–274. doi:10.1587/nolta.12.257

Pachiappan, K., Anitha, K., Pitchai, R., Sangeetha, S., Satyanarayana, T., & Boopathi, S. (2024). Intelligent Machines, IoT, and AI in Revolutionizing Agriculture for Water Processing. In *Handbook of Research on AI and ML for Intelligent Machines and Systems* (pp. 374–399). IGI Global.

Pitchai, R., Guru, K. V., Gandhi, J. N., Komala, C. R., Kumar, J. R. D., & Boopathi, S. (2024). Fog Computing-Integrated ML-Based Framework and Solutions for Intelligent Systems: Digital Healthcare Applications. In Technological Advancements in Data Processing for Next Generation Intelligent Systems (pp. 196–224). IGI Global. doi:10.4018/979-8-3693-0968-1.ch008

Prabhuswamy, M., Tripathi, R., Vijayakumar, M., Thulasimani, T., Sundharesalingam, P., & Sampath, B. (2024). A Study on the Complex Nature of Higher Education Leadership: An Innovative Approach. In *Challenges of Globalization and Inclusivity in Academic Research* (pp. 202–223). IGI Global. doi:10.4018/979-8-3693-1371-8.ch013

Pramila, P., Amudha, S., Saravanan, T., Sankar, S. R., Poongothai, E., & Boopathi, S. (2023). Design and Development of Robots for Medical Assistance: An Architectural Approach. In Contemporary Applications of Data Fusion for Advanced Healthcare Informatics (pp. 260–282). IGI Global.

Puranik, T. A., Shaik, N., Vankudoth, R., Kolhe, M. R., Yadav, N., & Boopathi, S. (2024). Study on Harmonizing Human-Robot (Drone) Collaboration: Navigating Seamless Interactions in Collaborative Environments. In Cybersecurity Issues and Challenges in the Drone Industry (pp. 1–26). IGI Global.

Rahamathunnisa, U., Sudhakar, K., Murugan, T. K., Thivaharan, S., Rajkumar, M., & Boopathi, S. (2023). Cloud Computing Principles for Optimizing Robot Task Offloading Processes. In *AI-Enabled Social Robotics in Human Care Services* (pp. 188–211). IGI Global. doi:10.4018/978-1-6684-8171-4.ch007

Rahamathunnisa, U., Sudhakar, K., Padhi, S., Bhattacharya, S., Shashibhushan, G., & Boopathi, S. (2024). Sustainable Energy Generation From Waste Water: IoT Integrated Technologies. In Adoption and Use of Technology Tools and Services by Economically Disadvantaged Communities: Implications for Growth and Sustainability (pp. 225–256). IGI Global.

Ravisankar, A., Sampath, B., & Asif, M. M. (2023). Economic Studies on Automobile Management: Working Capital and Investment Analysis. In Multidisciplinary Approaches to Organizational Governance During Health Crises (pp. 169–198). IGI Global.

Rebecca, B., Kumar, K. P. M., Padmini, S., Srivastava, B. K., Halder, S., & Boopathi, S. (2024). Convergence of Data Science-AI-Green Chemistry-Affordable Medicine: Transforming Drug Discovery. In *Handbook of Research on AI and ML for Intelligent Machines and Systems* (pp. 348–373). IGI Global.

Reddy, M. A., Gaurav, A., Ushasukhanya, S., Rao, V. C. S., Bhattacharya, S., & Boopathi, S. (2023). Bio-Medical Wastes Handling Strategies During the COVID-19 Pandemic. In Multidisciplinary Approaches to Organizational Governance During Health Crises (pp. 90–111). IGI Global. doi:10.4018/978-1-7998-9213-7.ch006

Revathi, S., Babu, M., Rajkumar, N., Meti, V. K. V., Kandavalli, S. R., & Boopathi, S. (2024). Unleashing the Future Potential of 4D Printing: Exploring Applications in Wearable Technology, Robotics, Energy, Transportation, and Fashion. In Human-Centered Approaches in Industry 5.0: Human-Machine Interaction, Virtual Reality Training, and Customer Sentiment Analysis (pp. 131–153). IGI Global.

Sangeetha, M., Kannan, S. R., Boopathi, S., Ramya, J., Ishrat, M., & Sabarinathan, G. (2023). Prediction of Fruit Texture Features Using Deep Learning Techniques. *2023 4th International Conference on Smart Electronics and Communication (ICOSEC)*, 762–768.

Sharma, D. M., Ramana, K. V., Jothilakshmi, R., Verma, R., Maheswari, B. U., & Boopathi, S. (2024). Integrating Generative AI Into K-12 Curriculums and Pedagogies in India: Opportunities and Challenges. *Facilitating Global Collaboration and Knowledge Sharing in Higher Education With Generative AI*, 133–161.

Sonia, R., Gupta, N., Manikandan, K., Hemalatha, R., Kumar, M. J., & Boopathi, S. (2024). Strengthening Security, Privacy, and Trust in Artificial Intelligence Drones for Smart Cities. In *Analyzing and Mitigating Security Risks in Cloud Computing* (pp. 214–242). IGI Global. doi:10.4018/979-8-3693-3249-8.ch011

Subha, S., Inbamalar, T., Komala, C., Suresh, L. R., Boopathi, S., & Alaskar, K. (2023). A Remote Health Care Monitoring system using internet of medical things (IoMT). *IEEE Explore*, 1–6.

Venkatasubramanian, V., Chitra, M., Sudha, R., Singh, V. P., Jefferson, K., & Boopathi, S. (2024). Examining the Impacts of Course Outcome Analysis in Indian Higher Education: Enhancing Educational Quality. In Challenges of Globalization and Inclusivity in Academic Research (pp. 124–145). IGI Global.

Venkateswaran, N., Kumar, S. S., Diwakar, G., Gnanasangeetha, D., & Boopathi, S. (2023). Synthetic Biology for Waste Water to Energy Conversion: IoT and AI Approaches. *Applications of Synthetic Biology in Health. Energy & Environment*, 360–384.

Venkateswaran, N., Vidhya, K., Ayyannan, M., Chavan, S. M., Sekar, K., & Boopathi, S. (2023). A Study on Smart Energy Management Framework Using Cloud Computing. In 5G, Artificial Intelligence, and Next Generation Internet of Things: Digital Innovation for Green and Sustainable Economies (pp. 189–212). IGI Global. doi:10.4018/978-1-6684-8634-4.ch009

Verma, R., Christiana, M. B. V., Maheswari, M., Srinivasan, V., Patro, P., Dari, S. S., & Boopathi, S. (2024). Intelligent Physarum Solver for Profit Maximization in Oligopolistic Supply Chain Networks. In *AI and Machine Learning Impacts in Intelligent Supply Chain* (pp. 156–179). IGI Global. doi:10.4018/979-8-3693-1347-3.ch011

Wilhelm, J., Petzoldt, C., Beinke, T., & Freitag, M. (2021). Review of digital twin-based interaction in smart manufacturing: Enabling cyber-physical systems for human-machine interaction. *International Journal of Computer Integrated Manufacturing*, *34*(10), 1031–1048. doi:10.1080/0951192X.2021.1963482

Chapter 8
Current and Future Research Directions

Himadri Sekhar Das

https://orcid.org/0000-0002-3509-3388
Haldia Institute of Technology, India

ABSTRACT

The union of machine intelligence (AI) and manlike robotics has brought about extraordinary progresses in miscellaneous rules, promising the concoction of an extreme-smart information technology. These branches investigate current research trends and future guidance in used AI and manlike electronics for the growth of the extreme-smart cyberspace. The authors argue key sciences, challenges, and potential uses in various fields to a degree healthcare, education, amusement, and manufacturing. Additionally, they investigate moral concerns and societal impacts guide the unification of AI and manlike science into information technology. This comprehensive review aims to support acumens into the developing countryside of AI and manlike robotics research, leading future endeavors towards achieving the thorough potential of the extreme-smart computer network.

1. INTRODUCTION

The brisk progresses in artificial intelligence (AI) and the study of computers have transformed miscellaneous aspects of human growth, varying from healthcare and production to entertainment and conveyance. One of the important frontiers concerning this union is the growth of ultra-smart web, place bright systems seamlessly communicate accompanying consumers and the environment. Applied AI and manlike science play important roles in forming this modern countryside,

DOI: 10.4018/979-8-3693-2399-1.ch008

offering creative answers to complex challenges and opening new potential for human-tool cooperation (Bajwa J et al, 2021). In this chapter, the journey an inclusive investigation of the current state and future research directions in used AI and manlike the study of computers for ultra-smart web. Outlining the foundational ideas and current progresses in these fields. Subsequently, into the unification of AI and electronics into web, examining the challenges and time that stand in this endeavor. Furthermore, go across the potential impacts of these sciences on institution, economy, and electronics countryside. Finally, outlined hopeful avenues for future test, conceiving the course of applied AI and manlike the study of computers in forming the ultra-smart information technology of later. In this framework, applied AI and manlike machine intelligence serve as the foundation of extreme-smart web, enabling independent administrative, adjusting behavior, and framework-knowledgeable interplay (Quinn TP et al, 2021). Through the integration of AI algorithms and done or made by machine podiums, computer network becomes infused accompanying intelligent capabilities, permissive it to see, reason, and act in real-occasion. For instance, independent instruments equipped accompanying AI-compelled traveling systems guide along route, often over water city streets, while manlike androids assist users in in essence atmospheres. Furthermore, the increase of Internet of Things (IoT) devices and sensors embellishes the neurological idea of cyberspace, permissive it to capture and resolve vast amounts of dossier from the substance (Topol EJ. 2019). This dossier-driven approach furthers predicting science of logical analysis, anomaly discovery, and growth, superior to more efficient support exercise and upgraded user knowledge. Moreover, edge estimating architectures scatter computational tasks, enabling faster reaction occasions and shortened latency in extreme-smart computer network.

2. OVERVIEW OF APPLIED AI AND HUMANOID ROBOTICS

Applied AI and manlike science represent two contemporary fields at the crossroads of machine intelligence (AI) and robotics, accompanying meaningful potential to impact various enterprises and facets of human existence. Here's an overview of each: Applied AI:

Definition: Applied AI includes the exercise of machine intelligence techniques to answer legitimate-world questions and embellish existent systems across differing rules. Techniques: Applied AI surrounds a range of techniques containing machine intelligence, natural language processing, calculating fantasy, and growth algorithms, among remainder of something (Sendak MP et al..2020) and (Haque A et al, 2020).

Applications: Applied AI finds requests in different fields such as healthcare, finance, conveyance, production, retail, and more.

Examples:

Healthcare: Diagnostic orders stimulate by AI, embodied treatment advice schemes, predicting analytics for patient consequences.

Finance: Fraud discovery systems, concerning mathematics business, department dealing with customers chatbots, risk management.

Transportation: Autonomous jeeps, traffic administration structures, predictive perpetuation for flotillas.

Manufacturing: Predictive maintenance of equipment, control of product quality through calculating vision, addition of result processes.

Retail: Recommendation plans, inventory administration, department dealing with customers automation.

Humanoid Robotics

Definition: Humanoid electronics focuses on crafty and construction robots that simulate and communicate accompanying humans somewhat, two together physically and behaviorally. Design: Humanoid machines are created to mimic human evolutions and appearance, accompanying facial characteristics like weaponry, legs, shape, and a head. Functionality: These androids are capable of operating tasks grazing from plain actions like ambulatory and greedy objects to complex interplays such as ideas and moving expression. Applications: Humanoid machines have requests in regions such as healthcare, instruction, pleasure, department dealing with customers, and research (Y. Tong et al,2024) and (Dezhen Xiong et al,2024) and (Yunjun Zheng et al,2024).

Examples: Assistive Robotics: Humanoid robots created to assist things with restrictions or retired things with tasks like attractive objects, providing notices, or accompaniment.

Education and Research: Humanoid robots secondhand in research labs and instructional institutions for learning human-android interplay, cognitive incident, and surveying AI algorithms. Entertainment: Humanoid androids employed in idea parks, exhibitions, and amusement venues for shared happenings and depictions.

Customer Service: Humanoid robots redistributed in sell surroundings or hospitality areas to support information, counseling, or amusement to consumers. In recent age, skilled has happened increasing unification betwixt applied AI and manlike machine intelligence, accompanying AI technologies reinforcing the potential of manlike robots and permissive more cosmopolitan interactions accompanying persons. This union holds the potential to revolutionize businesses to a degree healthcare, department dealing with customers, and education by forging smarter and capable electronic plans. However, moral considerations concerning the arrangement and use of AI-stimulate humanoid machines wait important, containing concerns about task dislocation, privacy, and the affect human friendly interplays.

2.1 Definition and Concepts

Applied Artificial Intelligence (AI) and Humanoid Robotics show contemporary fields at the intersection of data processing, architecture, and intelligent erudition. Applied AI involves the happening and exercise of astute systems to resolve experienced questions across differing domains (Demetris Vrontis et al,2022). These structures pretend human perception processes in the way that learning, interpretation, and logical to act tasks efficiently. In manlike electronics, the focus act building robots accompanying human-like presentations and functionalities. These androids are designed to communicate accompanying persons and their surroundings in a manner looking like human nature (Aleksander, I. et al,2017).

2.2 Historical Development

The growth of Applied AI and Humanoid Robotics can be tracked back to early attempts to constitute machines fit smart behavior. In the 1950s and 1960s, scientists started surveying the possibilities of AI through programs like the Logic Theorist and the General Problem Solver (Raj, R.et al,2022). These works designed the basis for later progresses in AI technology. In the field of science, early manlike machines like Shakey the Robot arose in the 1970s, showcasing elementary efficiencies in the way that mobility and understanding. Over the following decades, progresses in matters skill, computer concept, and pretended affecting animate nerve organs networks led to meaningful progress in manlike the study of computers (Russell, S et al.2022). The 21st of one hundred years witnessed a surge in interest and asset in two together Applied AI and Humanoid Robotics. Breakthroughs in machine intelligence, particularly deep knowledge, transformed AI capacities, permissive systems to realize exceptional levels of efficiency in tasks to degree image acknowledgment and machine intelligence. Concurrently, progresses in robotics fittings and program concreted the habit for the development of more advanced manlike robots. Early Robotics and AI: The fundamentals for used AI and manlike science began in the intervening-20th centennial accompanying the development of fundamental machine intelligence and AI algorithms (N. Prakash et al,2023). Early machines were generally used in technical scenes and were restricted in their potential. Advancements in AI: The advent of machine intelligence and affecting animate nerve organs networks in the late 20th centennial revolutionized AI, permissive androids to gain dossier and improve their act over period. This experienced to the development of more complex AI-stimulate machines fit performing complex tasks. Humanoid Robotics

Emergence: Humanoid the study of computers arose as a subfield meeting on creating androids accompanying human-like characteristics and campaigns. This involved progresses in fabrics erudition, mechanical metallurgy, and machine intelligence. Integration of AI and Robotics: As AI algorithms enhanced more effective, they were integrated into manlike androids, permissive ruling class to perceive and communicate accompanying the surroundings in more sophisticated habits. This unification managed to the growth of AI-driven independent manlike androids capable of operating tasks in miscellaneous rules (Bhushan, Tripti et al, 2024). Cyberspace and Connectivity: With the increase of the internet and mathematical electronics, computer network became an basic some human history. These supported new opportunities for mixing AI and manlike electronics into in essence environments, superior to the idea of extreme-smart cyberspace.

2.3 Key Components and Technologies

Applied AI and Humanoid Robotics depend a difference of elements and sciences to function effectively. In AI, key parts contain: Machine Learning Algorithms: These algorithms authorize systems to get or give an advantage dossier and help their accomplishment over time. Natural Language Processing (NLP): NLP methods admit schemes to understand and create human dialect, simplifying ideas between persons and machines. Computer Vision: Computer dream schemes enable machines to see and define optical information from their surroundings, permissive tasks in the way that object acknowledgment and navigation. Reinforcement Learning: Reinforcement education algorithms authorize wholes to learn through experimental approach, taking response from their atmosphere to improve their administrative talents.In manlike robotics, key elements involve: Actuators and Sensors: Actuators support machines with the talent to move and communicate accompanying their environment, while sensors allow ruling class to see and put oneself in the place of another stimuli (Xie, D.et al,2022). Humanoid Design: Humanoid androids are planned to simulate persons in appearance and action, expediting open interaction accompanying persons. Control Systems: Control orders rule the behavior of manlike androids, matching their movements and conduct. Human-Robot Interaction: Human-machine interplay sciences enable machines to write and cooperate with persons efficiently, easing uses in areas to a degree healthcare, instruction, and pleasure. Overall, Applied AI and Humanoid Robotics show exciting fields accompanying the potential to alter enterprises and society, contribution resolutions to complex questions and reinforcing human capabilities.

3. CURRENT STATE OF RESEARCH

In current age, the integration of machine intelligence (AI) in manlike robotics has signed meaningful progresses, leading to transformational impacts across miscellaneous rules. This integration has promoted the incident of robots that can see, determine, reason, and communicate with persons and their surroundings more naturally and cleverly. Here, we investigate the current state of research across various key areas:

3.1 AI Integration in Humanoid Robotics

The field of manlike the study of computers has visualized remarkable progress on account of AI unification. Research focuses on enhancing machine understanding, maneuverability, and dexterity through leading AI algorithms to a degree deep learning, support knowledge, and calculating vision. This allows machines to act complex tasks in unstructured atmospheres, looking like human capabilities more carefully. Integrating machine intelligence (AI) into manlike robotics shows a contemporary field with big associations for differing domains, containing manufacturing, healthcare, instruction, and entertainment. The union of AI and manlike robotics aims to generate creative structures capable of alert, interpretation, and communicating with their atmospheres and persons in a manner similar to human understanding and attitude (Doncieux, S.et al,2022). Figure 1 shows a example of AI Integration in Humanoid Robotics.

Figure 1. AI integration in humanoid robotics

Current and future research directions situated on sides are important for advancing the efficiencies of manlike androids and facilitating their unification into the extreme-smart information technology. Here, the outline some key research guidance's and their potential impact:

Machine Learning and Adaptive Control: Advancements in machine intelligence techniques, in the way that deep knowledge, support learning, and transformative algorithms, are important for enabling manlike androids to gain data and acclimate to active surroundings. Research in this area focuses on expanding algorithms for understanding, decision-making, and control, admitting machines to alone perform tasks and communicate accompanying humans carefully and efficiently.

Sensor Fusion and Perception: Humanoid androids require strong understanding schemes to understand their environment and communicate with objects and persons. Research in sensor mixture, containing vision, wisdom feeling, lidar, and tactile noticing, aims to improve androids' perceptual powers, permissive bureaucracy to perceive and define complex surroundings accurately. Natural Language Processing and Dialog Systems: Integrating machine intelligence (NLP) potential into manlike robots promotes logical communication 'tween persons and machines. Research focuses on developing talk structures that allow robots to learn and create human-like speech, define consumer commands, answer questions, and undertake meaningful discourses.

Human-Robot Interaction (HRI): Humanoid androids need to interact accompanying persons in culturally acceptable and instinctive habits. Research in HRI investigates human-like gestures, first expressions, and spoken hints to enhance androids' informative talents and foster unaffected and charming interactions accompanying consumers. Autonomous Navigation and Mobility: Autonomous guiding along route, often over water is critical for manlike machines to guide along route, often over water indoor and rustic atmospheres safely and capably (González-García et al, 2020). Research circumference focuses on cultivating algorithms for simultaneous localization and plan (SLAM), way planning, barrier eluding, and vital environment understanding, permissive machines to guide along route, often over water complex terrains autonomously. Cognitive Architectures and Emotion Understanding: To work efficiently in human-principal atmospheres, manlike robots need intelligent architectures worthy reasoning, preparation, and understanding human affections and purposes. Research in cognitive science aims to expand architectures stimulated by human cognition, permissive androids to exhibit adaptive and excitedly wise demeanor. Ethical and Social Implications: As humanoid machines enhance more and more integrated into institution, it is owned by address ethical and public associations had connection with their deployment. Research situated on sides investigate ethical concerns, solitude concerns, enlightening differences, and social agreement of manlike robots, guaranteeing trustworthy and ethical arrangement in the extreme-smart web (Ferrera, M et al.2018).

Multi-Robot Collaboration and Swarm Intelligence: Collaborative multi-robot wholes offer reinforced capabilities for carrying out complex tasks and operating in big surroundings. Research in swarm robotics focuses on evolving arrangement algorithms and ideas protocols for permissive logical collaboration and composite administrative with groups of humanoid androids. Augmented Reality and Virtual Reality Integration: Integrating improved reality (AR) and computer simulation (VR) sciences accompanying humanoid machine intelligence authorizes mesmerizing human-robot interplays and improved teleoperation capabilities. Research circumference investigate arrangements for fusing actual-globe sensor data accompanying in essence atmospheres, enabling androids to see and communicate with improved or fake surroundings.

Continual Learning and Lifelong Adaptation: Humanoid machines must steadily determine and adapt to developing atmospheres and user desires. Research in constant education and lifelong adjustment investigates algorithms for increasing by additions knowledge, knowledge memory, and familiarization to novel tasks and environments over comprehensive periods, guaranteeing androids' long-term independence and influence. By addressing these research guidances, chemists and engineers can drive novelties in AI and humanoid the study of computers, concreting the habit for the widespread unification of knowledgeable robots into the extreme-smart information technology and converting the way persons communicate with science from now on.

3.2 Human-Robot Interaction

Human-machine interplay (HRI) remains a critical region of research, intending to develop the seamless ideas and cooperation middle from two points humans and androids. Studies survey orders to improve robot understanding of human empathy and purposes, develop machine intelligence for productive ideas, and design instinctive interfaces for intuitive interplay (Admoni, H.et al, 2016). Additionally, research delves into guaranteeing the security and trustworthiness of machines close by physically to persons. Human-Robot Interaction (HRI) is a combining several branches of learning field that explores the design, exercise, and judgment of electronic systems that communicate accompanying persons. With the progress of Artificial Intelligence (AI) and humanoid the study of computers, HRI has supported meaningful progress, enabling machines to seamlessly mix into differing facets of human life. As we look towards the future, the union of used AI and manlike robotics is balanced to transform the idea of Ultra-Smart Cyberspace, place intelligent powers, containing machines, coexist and collude accompanying persons in mathematical and physical atmospheres (Ahn, H. S.et al,2019). Here are few current and future research guidance in this rule: Natural Language Understanding and Generation: Enhancing androids' strength to accept and generate human language is critical for direct human-robot ideas. Future research will devote effort to something cultivating AI algorithms that authorize robots to understand circumstances, conclude intentions, and counter suitably in discourses, permissive more natural and instinctive interplays.

Emotion Recognition and Expression: Emotion plays a essential duty in human ideas and accountable (Alghowinem, S.et al,2021). Future manlike robots will need expected outfitted accompanying leading emotion acknowledgment powers to see and respond to human concerning feelings and intuition states correctly. Furthermore, research will devote effort to something permissive robots to express concerns through spoken and non-spoken cues, supporting understanding interplays. Social Intelligence and Behavior Modeling: Humanoid androids must possess public understanding to guide along route, often over water complex social movement and obey public averages. Future research will focus on cultivating AI models that allow androids to understand friendly hints, accustom their act accordingly, and build complete friendships accompanying humans, promoting trust and aid (Aly, A.et al,2012).

Personalization and Adaptation: Every individual has singular desires, behaviors, and ideas styles. Future manlike androids will need to personalize their interplays established individual traits, past occurrences, and contextual clues. Research or in general area will devote effort to something developing AI algorithms that authorize androids to fit their management, communication style, and task killing approaches to suit individual weaknesses and requirements. Collaborative Problem-Solving and Task Execution: In Ultra-Smart Cyberspace, persons and machines will participate on differing tasks, ranging from household tasks to complicated situation-resolving activities. Future research will devote effort to something expanding AI algorithms that authorize machines to collaborate efficiently accompanying persons, share intentions, coordinate conduct, and donate intentionally to joint tasks, reinforcing overall task performance and adeptness (Andhare, P.et al,2016).

Ethical and Trustworthy AI: As androids enhance increasingly joined into human humankind, guaranteeing moral behavior and construction trust middle from two points persons and robots are principal. Future research will devote effort to something cultivating righteous AI frameworks, transparence means, and responsibility measures to ensure that androids obey righteous standard and societal standards, supporting trust, agreement, and adoption by persons. Physical and Virtual Embodiment: Humanoid machines can manifest in tangible form or as in essence avatars in digital atmospheres. Future research will survey the unification of physical and in essence manifestation, permissive logical transitions betwixt material and in essence interactions. Additionally, progresses in tele-presence electronics will allow detached operation of manlike androids, furthering human-robot interplays across terrestrial frontiers.

Long-Term Autonomy and Learning: Humanoid machines must exhibit long-term independence and steadily develop their capabilities through education and acclimatization (Ashok, K.et al,2022). Future research will devote effort to something evolving lifelong knowledge algorithms that authorize machines to acquire new abilities, readjust to changeful atmospheres, and autonomously boost their act over opportunity, reducing the need for manual attack and project.

The union of used AI and humanoid machine intelligence holds overwhelming promise for forming the future of Ultra-Smart Cyberspace. By discussing the aforementioned research guidances, we can solve the adequate potential of human-robot interplay, permissive machines to seamlessly merge into various facets of human existence and influence the creation of a brisker, completing, and all-embracing organization.

3.3. Applications in Healthcare

The unification of AI in manlike robotics has managed to meaningful breakthroughs in healthcare uses. Robots outfitted with AI can assist in patient care, restoration, and enucleation, improving the powers of healthcare professionals. These machines can monitor patient signs of life, determine embodied therapy, and even act sensitive surgical processes accompanying precision and effectiveness (Deo N et al,2023). Fig.2 AI and robotics has managed to meaningful breakthroughs in different healthcare uses.In the sphere of healthcare, the mixture of used AI and humanoid the study of computers holds huge promise for transforming healing practices and patient care within the circumstances of extreme-smart information technology. Here are few current and future research directions in this place field:

Figure 2. AI and robotics has managed to meaningful breakthroughs in different healthcare uses

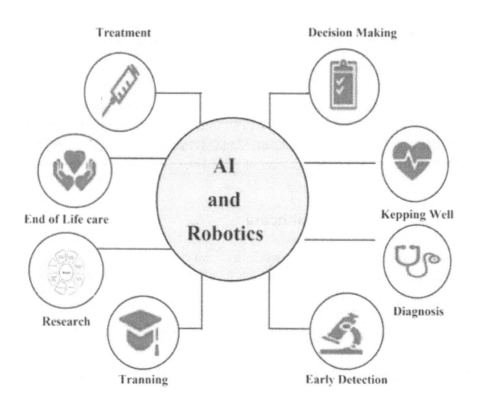

Robotic Surgery and Intervention: Applied AI joined accompanying manlike robotics allows more exact, minimally obtrusive surgeries and mediations. Future research aims to enhance made or done by a machine ability, absolute-period decision-making efficiencies, and haptic response orders to raise surgical outcomes (Ramalingam B et al, 2020).

Personalized Medicine and Treatment Planning: AI algorithms resolve far-flung amounts of patient dossier to embody treatment plans and call individual reactions to drug. Humanoid androids can assist in delivering tailor-made care, listening patient progress, and providing original-opportunity feedback to healthcare providers.

Remote Patient Monitoring and Telemedicine: Humanoid machines outfitted accompanying AI can be a part of companions for the aging or things accompanying never-ending conditions, providing friendship, warnings for cure, and listening vital signs. Research focuses on reconstructing the study of computers (NLP) wherewithal for persuasive communication between victims and androids.

Rehabilitation and Physical Therapy: Humanoid machines with AI-compelled adjusting education competencies can assist in rehabilitation exercises and material medicine gatherings. Future research aims to expand robots that can dynamically regulate exercise procedures established patient progress and response.

Diagnostic Imaging and Interpretation: AI algorithms integrated accompanying manlike electronics aid in the understanding of medical image, to a degree X-beams, MRIs, and CT scans (Wada K et al,2003). Research focuses on reconstructing the accuracy of representation study algorithms and cultivating androids capable of alone operating demonstrative processes. Drug Discovery and Development: AI algorithms assist in the discovery of novel drug aspirants and envision their efficiency and security profiles. Humanoid machines can mechanize extreme-throughput hide processes and accelerate drug incident timelines.

Healthcare Workforce Support: Humanoid androids outfitted accompanying AI can assist healthcare professionals by operating routine tasks, to a degree patient consumption, organizing appointments, and upholding photoelectric fitness records. Future research aims to embellish the collaboration middle from two points androids and human healthcare employees in dispassionate settings (Bera K et al,2019).

Ethical and Regulatory Considerations: As AI-compelled manlike androids enhance more integrated into healthcare plans, skilled is a need to address moral concerns had connection with patient privacy, consent, and the accountable use of AI sciences. Future research endure devote effort to something developing directions and supervisory foundations to guarantee the ethical arrangement of AI in healthcare. By concentrating on these research guidance's, the unification of applied AI and manlike electronics in healthcare can bring about the production of ultra-smart information technology surroundings that reinforce patient consequences, improve the adeptness of healthcare transmittal, and advance healing information and innovation.

3.4 Education and Training

AI-joined manlike androids are progressively being utilized in instructional backgrounds to reinforce education experiences. These machines comprise tutors, tutors, or shared learning helpers, providing to distinguished knowledge needs and promoting date and memory. Moreover, they specify experiential training in differing fields, to a degree prioritize, construction, and vocational abilities, fitting things for the trained workers of the future. Education and training in used AI and manlike machine intelligence for the extreme-smart cyberspace bear contain a blend of hypothetical information, practical abilities, and forward-thinking research methods. Here's an outline of what aforementioned instruction and training ability require: Foundational Concepts: Understanding the fundamental law of machine intelligence, including machine intelligence, deep knowledge, the study of computers, calculating vision, and electronics. Grasping the fundamentals of manlike machine intelligence, including kinematics, action, sensor unification, and control algorithms. Exploring the hypothetical supports of human cognition, attitude, and performance for plotting AI methods that interact seamlessly accompanying persons.

Advanced AI Techniques: Deep diving into leading machine intelligence techniques in the way that support education, fruitful adversarial networks (GANs), and transfer education. Exploring contemporary research in AI morality, justice, interpretability, and transparency to guarantee accountable AI arrangement in computer network. Understanding the nuances of AI cybersecurity and strength to ward off opposing attacks and guarantee the integrity of AI plans.

Humanoid Robotics Development: Hands-on occurrence in plotting, construction, and programming manlike androids, containing fittings components to a degree actuators, sensors, and idea plans. Learning the study of computers and frameworks usually secondhand in science happening, such as ROS (Robot Operating System) and Python. Experimenting accompanying imitation surroundings for expeditious prototyping and testing of electronic algorithms before legitimate-experience arrangement.

Cyberspace Integration: Understanding the unique challenges and moment bestowed by merging AI and manlike robotics into the extreme-smart web. Exploring integrative fields such as human-calculating interplay, computer simulation, improved reality, and Internet of Things (IoT) for conceiving mesmeric and instinctive interfaces. Studying the associations of AI and robotics on miscellaneous areas in the way that healthcare, instruction, entertainment, conveyance, and smart municipalities (Pradhan B et al, 2021).

Research Directions: Staying next to new research trends and breakthroughs in used AI and manlike machine intelligence through information reviews, academic conferences, and cooperation accompanying manufacturing spouses. Identifying emerging research guidances in the way that explicable AI, lasting learning, human-machine cooperation, swarm machine intelligence, and neuromorphic calculating. Cultivating critical thinking and logical abilities to address complex challenges in extreme-smart information technology development, in the way that solitude concerns, dossier governance, and social impact.

Practical Projects and Case Studies: Engaging in experiential projects and case studies that request AI and manlike robotics to palpable-globe questions in extreme-smart cyberspace, under the counseling of knowing instructors. Collaborating accompanying peers to tackle interdisciplinary challenges and influence various outlooks for creative solutions. Encouraging enterprise and change by promoting an progressive mindset and providing support for startups and development programs in AI and machine intelligence rules.

Continuous Learning and Adaptation: Emphasizing the significance of lifelong education and unending acclimatization to make even rapid progresses in science and progressing consumer needs. Encouraging participation in connected to the internet courses, mills, hackathons, and open-beginning societies to expand information and abilities further established classroom scenes. Promoting a civilization of interest, test, and collaboration to drive continuous change in used AI and manlike robotics for the extreme-smart web.By merging these details into education and preparation programs, things can gain the multidisciplinary expertise and research guile wanted to shape the future of used AI and manlike robotics in the extreme-smart computer network.

3.5 Entertainment and Social Interaction

Robots introduced accompanying AI is revolutionizing pleasure and friendly interplay rules. From interactive friends and individual helpers to entertainers and entertainers, AI-driven manlike machines offer novel happenings and amusement avenues. These machines interconnect consumers in funny interactions, description, and even psychological support, improving public experiences and improving comfort.

3.6 Industrial Automation

In modern backgrounds, AI-integrated manlike androids are streamlining production processes and growing efficiency. These machines can act repetitious tasks accompanying precision and constancy, accommodate to vital atmospheres, and collaborate accompanying human traders seamlessly. From congregation lines to management and maintenance, AI-compelled manlike machines are forceful innovation and output in miscellaneous mechanical areas.Industrial automation is having important renewals compelled by advancements in used machine intelligence (AI) and manlike machine intelligence (Allam Hamdan et al,2021). These technologies are forming the future of technical processes, permissive the invention of ultra-smart information technology surroundings.

Here are few current and future research guidances in this rule:

AI-Driven Predictive Maintenance: AI algorithms are being used to forecast supplies declines and perform perpetuation tasks before breakdowns happen. Future research will devote effort to something reconstructing the accuracy and dependability of predicting support plans by incorporating state-of-the-art machine intelligence methods and absolute-time sensor dossier (Benbya, Hind et al, 2020).

Collaborative Robots (Cobots): Humanoid androids are being planned to work alongside human peasants in industrial scenes, embellishing output and security. Research in this area will survey habits to raise the cooperation between persons and androids, containing human language interaction, expression acknowledgment, and joint in charge processes.

Autonomous Mobile Robots: Autonomous mobile androids are more and more being redistributed in warehouses and production facilities to mechanize material management tasks. Future research will devote effort to something cultivating advanced traveling and way preparation algorithms to authorize robots to run in vital and unorganized atmospheres safely and capably.

Robotic Vision and Perception: Vision methods are detracting for androids to perceive and learn their environment. Research situated on sides will devote effort to something developing AI algorithms for object acknowledgment, setting understanding, and geographical interpretation, enabling androids to act complex tasks to a degree object guidance and assembly.

Human-Robot Interaction (HRI): As machines enhance more governing in mechanical environments, research into HRI will enhance more main. Future research will investigate ways to design machines that can efficiently write and collude with human employees, allowing for possibility friendly and enlightening factors (Kiran Jot Singh et al,2021). Edge Computing real-Time Control: Edge calculating electronics are being leveraged to act real-occasion data conversion and control in industrialized industrialization systems (Bayar, A et al. 2023). Future research will devote effort to something expanding edge AI algorithms that can develop support utilization, underrate abeyance, and guarantee strength in dynamic technical surroundings.

Explainable AI (XAI) for Safety and Transparency: As AI algorithms are progressively joined into industrial industrialization structures, guaranteeing their security and transparency enhances principal. Future research will investigate methods for making AI algorithms more interpretable and see-through, permissive human controllers to accept and trust their decisions.

Digital Twins and Simulation: Digital identical twins are in essence replicas of tangible property or processes that can be secondhand for imitation, growth, and predicting analytics. Future research will devote effort to something evolving more refined mathematical twin models that incorporate AI and machine intelligence methods to specify correct predictions and observations into modern processes.

Cybersecurity for Industrial AI Systems: As mechanical mechanization systems enhance completing and dependent on AI algorithms, they more become more ready to computerized attacks (Czeczot, G et al.2023). Future research will devote effort to something cultivating robust cybersecurity resolutions to assure industrialized AI plans from threats in the way that malware, dossier breaches, and sabotage.

Ethical and Societal Implications: Finally, research in mechanical industrialization must also address the righteous and pertaining to society suggestions of AI and manlike robotics. Future research will survey issues to a degree task dislocation, privacy concerns, and the impact of mechanization on human health, meaning to cultivate policies and directions that guarantee the accountable arrangement of these technologies. In conclusion, the union of used AI and manlike electronics is driving accelerated progresses in technical computerization, paving the habit for the growth of extreme-smart computer network environments.

3.7 Challenges and Limitations

Despite important progresses, challenges and limitations carry on the unification of AI in manlike electronics. These include guaranteeing strength and dependability of AI algorithms, calling ethical and security concerns, checking biases hesitation-making processes, and supporting acceptance and trust between consumers. Additionally, scalability and cost-influence wait key considerations for extensive endorsement across various requests and industries.

4. FUTURE RESEARCH DIRECTIONS

The swift progresses in artificial intelligence (AI) and electronics have thrown technology into new fields, hopeful life-changing impacts across various areas. In this discourse, we investigate future research directions in AI and science, emphasize key fields such as progresses in AI algorithms, reinforcing human-robot cooperation, embodied healthcare help, immersive knowledge surroundings, and emotional wit in ape, ethical concerns, bias alleviation, and tenable development.

4.1. *Advancements in AI Algorithms*: The occupation of more cultured AI algorithms remains a main focus for analysts. With the arrival of deep learning and affecting animate nerve organs networks, skilled has been unusual progress in tasks like figure acknowledgment, natural language processing, and independent accountable. However, challenges persist, specifically in expanding algorithms that can statement across domains, get or give an advantage restricted data, and readjust to vital atmospheres. Future research will likely explore novel architectures, to a degree neuro-representative approaches, reinforcement knowledge accompanying survey, and transfer learning example, to overcome these restraints and push the boundaries of AI wherewithal.

4.2. *Enhancing Human-Robot Collaboration*: As androids enhance increasingly joined into miscellaneous aspects of association, improving their ability to conspire efficiently accompanying humans is superior. Future research will devote effort to something developing instinctive interfaces, ideas codes, and shared independence methods to facilitate smooth interplay middle from two points humans and androids. Collaborative machines (cobots) equipped accompanying progressive understanding capabilities and adjusting acts will enable more reliable and more creative participations in domains varying from production and healthcare to household assistance and accident answer.

4.3. *Personalized Healthcare Assistance*: The crossroads of AI and healthcare holds tremendous promise for embodied healing assistance. By leveraging enormous amounts of patient dossier, containing genomic profiles, healing histories, and original-time physiologic versification, AI-powered structures can offer tailor-made situation recommendations, predicting disease, and proactive invasions. Future research will investigate leading machine learning models, in the way that allied learning and fruitful opposing networks, to address solitude concerns while maximizing the utility of healthcare dossier. Additionally, the unification of wearable sensors, IoT devices, and telehealth manifestos will authorize unending monitoring and embodied attacks, revolutionizing patient care transmittal.

4.4. *Immersive Learning Environments*: Immersive knowledge atmospheres, facilitated by computer simulation (VR) and improved reality (AR) electronics, hold huge potential to mutate education and preparation example (D'Urso, F.et al,2023). Future research will focus on cultivating AI-compelled content production, adaptive knowledge algorithms, and common simulations to create embodied and charming instructional experiences. By leveraging machine intelligence and calculating vision methods, in essence tutors can provide actual-period response, adapt education matters to individual preferences, and promote cooperative logical skills. Moreover, mesmeric education environments can balance approach to instruction by overcoming terrestrial hurdles and accommodating different knowledge styles.

4.5. *Emotional Intelligence in Humanoids*: The unification of emotional judgment into manlike robots shows a important boundary in AI research. Future advancements will aim to infuse machines with the talent to see, believe, and respond to human fervors efficiently. This involves evolving multimodal believing proficiencies, sentiment study algorithms, and concerning feelings and intuition computing foundations to define subtle clues in the way that first expressions, accent, and nonverbal communication. Humanoids equipped accompanying heated intellect can enhance human-machine interplay in contexts grazing from department dealing with customers and remedy to companionship and public support, supporting deeper rapport and understanding.

4.6. *Ethical Considerations and Bias Mitigation*: As AI sciences enhance increasingly extensive, sending ethical concerns and diminishing concerning manipulation of numbers biases is critical. Future research will devote effort to something expanding transparent and liable AI schemes that maintain principles of justice, transparence, and accountability. This contains mixing righteous frameworks into AI growth pipelines, advancing diversity and inclusivity in dataset group and model preparation, and implementing systems for bias discovery and alleviation. Moreover, interdisciplinary cooperation betwixt computer physicists, ethicists, policymakers, and shareholders will be owned by navigate complex righteous crises and ensure accountable AI arrangement.

4.7. *Sustainable Development and Energy Efficiency*: The occupation of sustainable incident and strength efficiency shows a important challenge for AI and the study of computers research (Dian M. Grueneich,2015). Future endeavors will aim to harness AI technologies to correct property utilization, diminish tangible impact, and advance sustainable practices across miscellaneous rules. This includes expanding AI-compelled resolutions for energy administration, smart foundation, and environmental monitoring to underrate element footmark and enhance elasticity to temperature change. Additionally, research efforts will investigate the unification of renewable energy beginnings, in the way that cosmic and wind power, accompanying AI-authorized smart grids and energy depository wholes to speed the transition towards a bearable and flexible future.

5. APPLICATIONS IN ULTRA-SMART CYBERSPACE

5.1. Smart Homes and IoT Integration: In extreme-smart computer network, smart homes enhance more joined and smart, with IoT tools seamlessly writing and matching to optimize strength custom, embellish security, and support embodied occurrences for inhabitants (Fig.3). Advanced AI algorithms survive miscellaneous designs and systems, in the way that ignition, warming, appliances, and pleasure, established consumer preferences, tendencies, and incidental environments. Research in applied AI and manlike machine intelligence for extreme-smart cyberspace, specifically in the framework of smart families and IoT integration, spans a roomy array of fields, accompanying both current and future guidance being of principal significance. Here are some key research guidances and concerns:

Context-Awareness and Personalization: Future research will likely devote effort to something developing AI wholes that can appreciate and suit to the preferences, tendencies, and needs of individual consumers inside smart homes (Zaidan, Aws Alaa, et al., 2018). This includes engaging leading machine learning algorithms to draw and resolve dossier from various sensors and IoT maneuvers to conclude consumer context and supply embodied duties and recommendations. Multi-modal Interaction: There is a increasing interest in evolving everyday and intuitive habits for persons to communicate with smart home plans and manlike androids. Research in this area concede possibility include merging speech acknowledgment, nod acknowledgment, facial verbalization study, and different modalities to authorize smooth ideas and interaction between persons and AI-compelled devices (Ghayvat, Hemant, et al.,2015).

Autonomous Decision Making: As smart home surroundings enhance more complex with a large group of pertain instruments and systems, skilled is a need for AI algorithms worthy independent decision-making. This contains evolving AI methods that can prioritize and kill tasks, survive strength usage, improve reserve distribution, and ensure security and protection inside the smart home environment.

Privacy and Security: Ensuring the solitude and safety of consumer data inside smart families is of maximum importance. Future research will devote effort to something cultivating healthy AI-driven answers for dossier encryption, approach control, anomaly discovery, and danger alleviation to safeguard sensitive facts and save against computerized-attacks and unauthorized approach.

Interoperability and Standardization: With the increase of IoT devices from differing manufacturers, skilled is a urgent need for interoperability standards and obligations to allow smooth communication and unification between assorted devices. Research works will likely devote effort to something expanding AI-driven resolutions for ploy finding, configuration, and interoperability administration inside smart home environments.

Energy Efficiency and Sustainability: As the demand for energy-effective and tenable sciences continues to evolve, research situated on sides will devote effort to something developing AI-compelled answers for optimizing strength usage, foreseeing strength use patterns, and integrating energy from undepletable source beginnings inside smart home environments to underrate material impact.

Robotic Assistants and Companions: Humanoid androids equipped accompanying AI potential have the potential to symbolize personal helpers and helpers inside smart homes. Future research will devote effort to something evolving philosophically intelligent machines fit understanding human despairs, preferences, and purposes, and providing embodied help and companionship to consumers (Stojkoska, Biljana L et al, 2017).

Ethical and Social Implications: As AI-compelled sciences become more extensive inside smart houses, there is a need to address righteous and public suggestions such as concerning mathematics bias, task dislocation, and the impact on human connections. Future research will likely investigate foundations for ethical AI design, managing, and administrative to guarantee that AI technologies benefit organization all at once.

Figure 3. Smart homes and IoT integration

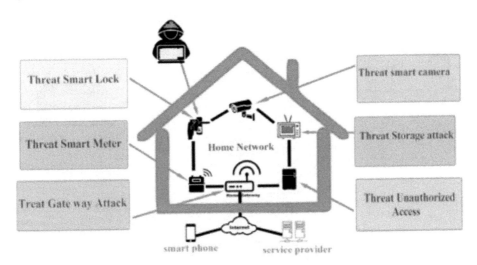

5.2. Virtual Assistants and Intelligent Agents: Virtual helpers and intelligent powers develop to become full of enthusiasm and expectant, leveraging vast amounts of dossier and cosmopolitan algorithms to understand and predict consumer needs and preferences correctly (Lamontagne, Luc, et al.2014). They offer embodied recommendations, mechanize routine tasks, and support valuable insights and help across miscellaneous domains, containing output, entertainment, buying, and healthcare.

5.3. Autonomous Vehicles and Transportation: In extreme-smart cyberspace, independent boats and transportation schemes are seamlessly pertain, communicating accompanying each one and with foundation components to optimize traffic flow, improve security, and minimize blockage (Bagloee, Saeed Asadi, et al.2016). Advanced AI algorithms allow vehicles to guide along route, often over water complex atmospheres, make real-period determinations, and adapt to changeful environments, leading to more effective, handy, and sustainable conveyance networks. Figure 4.shows the Aspects of Cyber Security in Autonomous and Connected Vehicles.

Figure 4. Aspects of cyber security in autonomous and connected vehicles

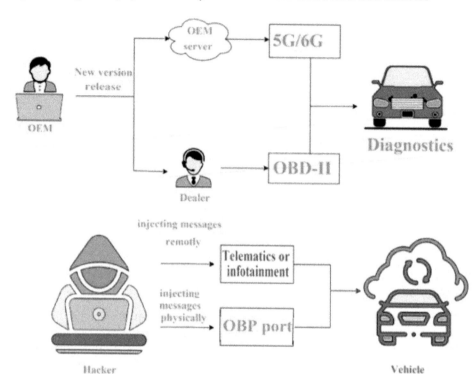

5.4. Cybersecurity and Threat Detection: Cyber-security enhances increasingly fault-finding in ultra-smart information technology, accompanying advanced AI-stimulate forms and techniques working to discover, prevent, and put oneself in the place of another high-tech threats efficiently. AI algorithms resolve vast amounts of dossier to recognize anomalies, envision potential attacks, and mechanize responses, reinforcing the elasticity and security of mathematical methods, networks, and infrastructure. Fig.5. shows the different types of Applications of cybersecurity.

Figure 5. Applications of different types of applications of cybersecurity

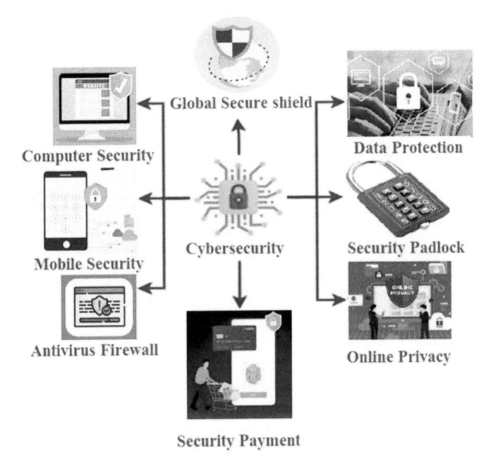

5.5. **Augmented Reality and Virtual Reality:** Augmented phenomenon (AR) and computer simulation (VR) technologies are joined into differing aspects of extreme-smart computer network, transforming in what way or manner community interact accompanying the mathematical world (Carlos Flavián et al,2019). AR overlays mathematical facts onto the tangible atmosphere, enhancing output, instruction, training, and amusement occurrences. VR creates mesmeric in essence environments for simulations, wager, public interactions, and detached collaboration, clouding foul line between the tangible and mathematical realms.

5.6. **Smart Cities and Infrastructure:** Smart capitals influence advanced sciences and dossier-driven visions to reinforce the efficiency, sustainability, and livability of city surroundings. In ultra-smart web, pertain sensors, actuators, and AI systems correct differing aspects of city foundation, containing energy, conveyance, waste administration, public safety, and city preparation (Gracias, J.S.et al, 2023). Real-time dossier science of logical analysis enable city administrators to form informed conclusions, increase resource distribution, and return promptly to arising challenges and time. Figure 6 shows the infrastructure of Smart Cities.

Figure 6. Smart cities and infrastructure

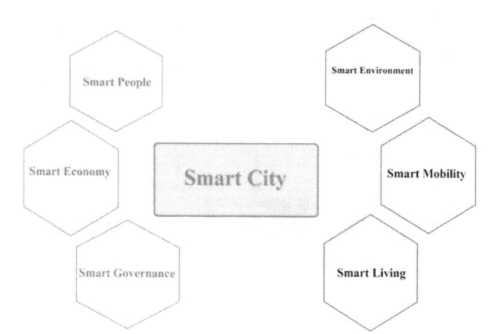

6. ETHICAL AND SOCIETAL IMPLICATIONS

Navigating the Ethical and Societal Implications of Advanced Technologies Introduction As electronics resumes to evolve at a speedy pace, it influences outward a myriad of moral and societal suggestions that warrant painstaking concern. From privacy concerns to task dislocation, bias, and supervisory foundations, the impact of advanced sciences filters miscellaneous facets of our lives. In this discourse, investigate these key suggestions, surveying their meaning and potential ramifications (Kendal E. Ethical, 2022).

6.1. Privacy Concerns: Privacy concerns have enhance progressively famous in the digital age, infuriate apiece conception of advanced sciences to a degree machine intelligence (AI) and considerable data science of logical analysis. These electronics have the ability to collect, resolve, and take advantage of far-reaching amounts of private data, lifting questions about the guardianship of individual solitude rights. From facial acknowledgment structures to dossier excavating algorithms, the potential for intrusive following and dossier using is substantial. Moreover, the com modification of private dossier by type of educational institution associations further amplifies solitude risks, as proved by abundant data breaches and debates. Addressing solitude concerns demands healthy regulatory foundations that supply instructions dossier protection, transparence, and individual consent while promoting change and concerning details advancement.

6.2. Job Displacement and Economic Impact: The coming of mechanization and AI-compelled technologies has started concerns about task dislocation and its more extensive business-related consequences. While computerization has the potential to streamline processes, boost output, and forge new task opportunities, it likewise warns established job sectors and infuriates pay prejudice. The displacement of reduced-skillful laborers by machines and algorithms emphasizes the need for reskilling and upskilling initiatives to gear things accompanying the competencies necessary for the tasks of the future. Additionally, policymakers must implement measures to check the antagonistic effects of industrialization on exposed peoples and ensure a just change to a technologically-compelled saving.

6.3. Human-Robot Relationships: The unification of robots and AI wholes into differing rules raises intriguing questions about human-machine friendships and social dynamics. From duty androids in healthcare to independent boats on the roads, humans are more communicating accompanying intelligent machines in various circumstances. As these interplays enhance more commonplace, it is crucial to analyze the moral suggestions of human-robot connections, containing issues of trust, responsibility, and understanding. Moreover, ensuring that AI methods exhibit righteous performance and adhere to social averages is necessary for promoting positive human-machine interplays and underrating the risk of unintended results (Benston, S.,2022).

6.4. Bias and Discrimination: The extensive attendance of bias and bias in AI algorithms poses significant moral challenges in the dimension of progressive technologies. Machine learning models prepared on partial datasets can maintain and even infuriate existing social prejudices, superior to discriminatory effects in extents in the way that hiring, accommodating, and police officers. Addressing concerning manipulation of numbers bias demands a multi-faceted approach including dossier variety, algorithmic transparence, and righteous supervision. Furthermore, advancing diversity and inclusivity inside the type of educational institution manufacturing is essential for mitigating bias and promoting impartial effects in AI-compelled systems.

6.5. Regulatory Frameworks and Governance: Effective supervisory foundations and government mechanisms are essential for directing the righteous and pertaining to society suggestions of advanced electronics. However, the speedy pace of mechanics innovation frequently outpaces the happening of appropriate tactics and regulations, superior to breach in care and accountability. Establishing healthy supervisory foundations that affect a balance between change and moral concerns is paramount. This requires cooperation middle from two points policymakers, technologists, ethicists, and other shareholders to predict arising challenges and conceive adaptive government makeups. Moreover, promoting international assistance and uniformity exertions can expedite harmonized approaches to technology requirement across borders.

In conclusion, guiding along route, often over water the moral and societal suggestions of leading sciences demands a concerted work to address key concerns to a degree solitude, job dislocation, human-machine friendships, bias, and supervisory governance. By supporting talk, advancing ethical best practices, and accomplishing full of enthusiasm tactics, we can harness the transformational potential of technology while ensuring the principles and rights of things and society loose. Only through composite date and prudence can we ensure that progressive sciences comprise forces for positive change in our more pertain realm.

7. FUTURE PROSPECTS AND RESEARCH

In the investigation of Applied AI and Humanoid Robotics for the Ultra-Smart Cyberspace, various indispensable content have emerged. Firstly, the unification of AI into miscellaneous facets of our lives has once started to transform labors, from healthcare to conveyance. Humanoid electronics, accompanying their capability to communicate and assist humans in tangible tasks, are flattering progressively advanced and intelligent. However, challenges such as righteous concerns, security concerns, and the need for strong foundation persist (Naneva, S.et al, 2020). Here argued the significance of integrative cooperation middle from two points' experts in AI, science, medicine, study of humans and their culture, and morality to address these challenges completely. Moreover, progresses in natural language processing, calculating dream and machine intelligence algorithms have concreted the habit for more intuitive and compliant manlike androids. Furthermore, the idea of the Ultra-Smart Cyberspace shows a future where AI and machine intelligence seamlessly merge into our day-to-day lives, embellishing output, efficiency, and condition of history (Rasekhipour, Y et al, 2017). This concept makes necessary continuous research and development exertions to accomplish allure complete potential while guaranteeing inclusivity, approachability, and ethical use of electronics. Future Prospects Looking advanced, the future of Applied AI and Humanoid Robotics holds huge promise. In the domain of AI, progresses in deep learning, support education, and allied education will allow more autonomous and creative structures. Humanoid machines will enhance more and more lifelike, accompanying upgraded olfactory proficiencies, aptitude, and changeability to diverse atmospheres. One inspiring prospect is the union of AI and the study of computers in the healthcare subdivision. Humanoid robots outfitted accompanying AI algorithms can assist healthcare artists in tasks varying from patient choose medical disease. These machines can determine friendship to the aging and individuals accompanying disadvantages, relieveing isolation and reconstructing insane well-being (Yurtsever, E et al, 2020). Moreover, as we progress towards the Ultra-Smart Cyberspace, embodied AI helpers will enhance ever-present, expecting our needs and preferences across miscellaneous rules. These helpers will seamlessly merge into our birthplaces, workplaces, and public spaces, improving availability and effectiveness. Additionally, research into moral AI and machine intelligence will remain superior. As these electronics enhance more extensive, guaranteeing transparency, responsibility, and justice in their in charge processes will be critical. Moreover, healthy high-tech security measures must be executed to safeguard against potential dangers and exposures (Alatise, M.B et al,2020). Recommendations for Further Research To further advance the field of Applied AI and Humanoid Robotics for the Ultra-Smart Cyberspace, various streets for research merit attention: Ethical Frameworks: Develop inclusive moral foundations for the design, arrangement, and organizing of AI and robotics electronics. Consideration concede possibility take to issues in the way that bias alleviation, privacy guardianship, and concerning manipulation of numbers transparence. Human-Robot Interaction: Investigate game plans to reinforce the ease and intuitiveness of human-android interplay. This involves research on token acknowledgment, emotion understanding, and adjusting ideas methods. Robustness and Safety: Explore procedures to embellish the robustness and security of manlike machines operating in active surroundings. This includes research on accident eluding, sin resistance, and danger reaction mechanisms. Social Impact: Assess the public impact of AI and manlike electronics on things, societies, and society loose. This contains examining the suggestions for hiring, education, and public similarity. Interdisciplinary Collaboration: Foster integrative cooperation middle from two points researchers, experts, policymakers, and collaborators from different rules (Saike, J et al,2022). This cooperative approach is essential for discussing complex challenges and ensuring the accountable growth and arrangement of AI and machine intelligence sciences. In conclusion, the pursuit of Applied AI and Humanoid Robotics for the Ultra-Smart Cyberspace holds excellent promise for transforming differing facets of our lives. By discussing key challenges, embracing multidisciplinary cooperation, and prioritizing righteous concerns, we can harness the thorough potential of these technologies to design a more all-encompassing, adept, and tenable future. Future Directions: Embodied AI: Future research will likely devote effort to something further integrating AI accompanying material manlike carcasses, permissive machines to embody acumen and communicate accompanying the substance more seamlessly. Human-Machine Integration: There will be raised emphasis on reinforcing the unification betwixt persons and machines in computer network, potentially chief to new example of human improving and cooperative connections with AI-compelled androids. Ethical and Societal Implications: As extreme-smart web enhances a truth, there will have influence dispute and research on the moral and social associations of these technologies, containing issues had connection with task dislocation, solitude, and autonomy. Biologically Inspired Robotics: Researchers concede possibility draw stimulus from physical science to evolve machines with more human-like proficiencies, to a degree bendable and flexible parties, in addition to systems that can self-repair and self-copy. Global Collaboration: Given the integrative character of research in used AI and manlike robotics for extreme-smart computer network, future guidance concede possibility include increased cooperation between scientists, engineers, policymakers, and ethicists from about the experience to address the complex challenges and opportunities bestowed by these electronics.

8. CONCLUSION

The investigation of used AI and manlike electronics within the domain of extreme-smart information technology offers a hopeful path for advancing science's capacities to do benevolence. Through this combination, several key observations arise, peeling arrive two together current achievements and future guidance. The crossroads of used AI and manlike the study of computers within extreme-smart information technology shows a boundary of novelty accompanying profound associations for institution. By dealing with a human-main, fairly mindful approach and promoting multidisciplinary cooperation, scientists can solve new possibilities and shape a future place electronics embellishes human potential and promotes all-embracing prosperity.

REFERENCES

Admoni, H., & Scassellati, B. (2016). Social eye gaze in human-robot interaction: A review. *Journal of Human-Robot Interaction*, 6(1), 25–63. doi:10.5898/JHRI.6.1.Admoni

Ahn, H. S. (2019). Hospital receptionist robot v2: design for enhancing verbal interaction with social skills. *2019 28th IEEE International Conference on Robot and Human Interactive Communication (RO-MAN)*, 1–6. 10.1109/RO-MAN46459.2019.8956300

Alatise, M.B. (2020). A Review on Challenges of Autonomous Mobile Robot and Sensor Fusion Methods. *IEEE Access, 8*, 39830–39846.

Aleksander, I. (2017). Partners of humans: A realistic assessment of the role of robots in the foreseeable future. *Journal of Information Technology, 32*(1), 1–9. doi:10.1057/s41265-016-0032-4

Alghowinem, S. (2021). Beyond the words: analysis and detection of self-disclosure behavior during robot positive psychology interaction. *2021 16th IEEE International Conference on Automatic Face and Gesture Recognition (FG 2021)*, 1–8. 10.1109/FG52635.2021.9666969

Allam Hamdan. (2021). The Relationship Between Intellectual Capital in the Fourth Industrial Revolution and Firm Performance in Jordan. *Studies in Computational Intelligence, 935*. https://www.springer.com/series/7092

Aly, A. (2012). Prosody-driven robot arm gestures generation in human-robot interaction. *Proceedings of the seventh annual ACM/IEEE international conference on Human-Robot Interaction*, 257–258. 10.1145/2157689.2157783

Andhare, P. (2016). Pick and place industrial robot controller with computer vision. *2016 International Conference on Computing Communication Control and automation (ICCUBEA)*, 1–4. 10.1109/ICCUBEA.2016.7860048

Ashok, K. (2022). Collaborative analysis of audio-visual speech synthesis with sensor measurements for regulating human-robot interaction. *Int. J. Syst. Assur. Eng. Manag.*, 1–8. . doi:10.1007/s13198-022-01709-y

Bagloee. (2016). Autonomous vehicles: challenges, opportunities, and future implications for transportation policies. *Journal of Modern Transportation, 24*, 284-303.

Bajwa, J., Munir, U., Nori, A., & Williams, B. (2021). Artificial intelligence in healthcare: Transforming the practice of medicine. *Future Healthcare Journal, 8*(2), 188–194. doi:10.7861/fhj.2021-0095 PMID:34286183

Bayar, A. (2023). Edge Computing Applications in Industrial IoT: A Literature Review. In J. Á. Bañares, J. Altmann, O. Agmon Ben-Yehuda, K. Djemame, V. Stankovski, & B. Tuffin (Eds.), Lecture Notes in Computer Science: Vol. 13430. *Economics of Grids, Clouds, Systems, and Services. GECON 2022.* Springer. doi:10.1007/978-3-031-29315-3_11

Benbya, H. (2020) Artificial Intelligence in Organizations: Current State and Future Opportunities. *MIS Quarterly Executive, 19*(4). doi:10.2139/ssrn.3741983

Benston, S. (2022). Walking a fine germline: Synthesizing public opinion and legal precedent to develop policy recommendations for heritable gene-editing. *Journal of Bioethical Inquiry, 19*(3), 421–431. Advance online publication. doi:10.1007/s11673-022-10186-8 PMID:35438443

Bera, K., Schalper, K. A., Rimm, D. L., Velcheti, V., & Madabhushi, A. (2019). Artificial intelligence in digital pathology - new tools for diagnosis and precision oncology. *Nature Reviews. Clinical Oncology, 16*(11), 703–715. doi:10.1038/s41571-019-0252-y PMID:31399699

Bhushan, T. (2024). Artificial Intelligence, Cyberspace and International Law. *Indonesian Journal of International Law, 21*(2), 3. https://scholarhub.ui.ac.id/ijil/vol21/iss2/3

Carlos Flavián. (2019). The impact of virtual, augmented and mixed reality technologies on the customer experience. *Journal of Business Research, 100*, 547-560. doi:10.1016/j.jbusres.2018.10.050

Czeczot, G., Rojek, I., Mikołajewski, D., & Sangho, B. (2023). AI in IIoT Management of Cybersecurity for Industry 4.0 and Industry 5.0 Purposes. *Electronics (Basel), 2023*(12), 3800. doi:10.3390/electronics12183800

D'UrsoF. (2023) Revolutionizing medical education: the impact of Virtual and Mixed Reality on training and skill acquisition. Preprints 2023, 2023121901. https://doi.org/ doi:10.20944/preprints202312.1901.v2

Deo, N., & Anjankar, A. (2023, May 23). Artificial Intelligence With Robotics in Healthcare: A Narrative Review of Its Viability in India. *Cureus, 15*(5), e39416. doi:10.7759/cureus.39416 PMID:37362504

Dian, M. G. (2015). The Next Level of Energy Efficiency: The Five Challenges Ahead. *The Electricity Journal, 28*(7), 44-56. doi:10.1016/j.tej.2015.07.001

Doncieux, S., Chatila, R., Straube, S., & Kirchner, F. (2022). Human-centered AI and robotics. *AI Perspect, 4*(1), 1. doi:10.1186/s42467-021-00014-x

Ethical, K. E. (2022, September). Legal and Social Implications of Emerging Technology (ELSIET) Symposium. *Journal of Bioethical Inquiry, 19*(3), 363–370. doi:10.1007/s11673-022-10197-5 PMID:35749026

Ferrera, M. (2018). The Aqualoc Dataset: Towards Real-Time Underwater Localization from a Visual-Inertial-Pressure Acquisition System. arXiv 2018, arXiv:1809.07076.

Ghayvat, H., Mukhopadhyay, S. C., Liu, J., Babu, A., Elahi, E., & Gui, X. (2015). Internet of Things for smart homes and buildings: Opportunities and Challenges. *Journal of Telecommunications and the Digital Economy, 3*(4), 33–47. doi:10.18080/jtde.v3n4.23

González-García, Gómez-Espinosa, A., Cuan-Urquizo, E., García-Valdovinos, L. G., Salgado-Jiménez, T., & Cabello, J. A. E. (2020). Autonomous Underwater Vehicles: Localization, Navigation, and Communication for Collaborative Missions. *Applied Sciences (Basel, Switzerland), 2020*(10), 1256. doi:10.3390/app10041256

Gracias, J. S., Parnell, G. S., Specking, E., Pohl, E. A., & Buchanan, R. (2023). Smart Cities-A Structured Literature Review. *Smart Cities, 2023*(6), 1719–1743. doi:10.3390/smartcities6040080

Haque, A., Milstein, A., & Fei-Fei, L. (2020). Illuminating the dark spaces of healthcare with ambient intelligence. *Nature, 585*(7824), 193–202. doi:10.1038/s41586-020-2669-y PMID:32908264

Kiran Jot Singh. (2021). All about human-robot interaction. In Cognitive Data Science in Sustainable Computing, Cognitive Computing for Human-Robot Interaction. Academic Press. doi:10.1016/B978-0-323-85769-7.00010-0

Lamontagne, L. (2014). A framework for building adaptive intelligent virtual assistants. *Artificial Intelligence and Applications (Commerce, Calif.), 10.*

Naneva, S., Sarda Gou, M., Webb, T. L., & Prescott, T. J. (2020). A Systematic Review of Attitudes, Anxiety, Acceptance, and Trust towards social Robots. *International Journal of Social Robotics, 2020*(12), 1179–1201. doi:10.1007/s12369-020-00659-4

Potter, Oluwaseyi, & Olaoye. (2024). Advances in Artificial Intelligence for Robotics. *Journal of Robotic Systems.*

Pradhan, B., Bharti, D., Chakravarty, S., Ray, S. S., Voinova, V. V., Bonartsev, A. P., & Pal, K. (2021). Internet of things and robotics in transforming current-day healthcare services. *Journal of Healthcare Engineering, 2021,* 9999504. doi:10.1155/2021/9999504 PMID:34104368

Prakash, N. (2023). Merging Minds and Machines: The Role of Advancing AI in Robotics. EAI Endorsed Trans IoT, 10.

Quinn, T. P., Senadeera, M., Jacobs, S., Coghlan, S., & Le, V. (2021). Trust and medical AI: The challenges we face and the expertise needed to overcome them. *Journal of the American Medical Informatics Association : JAMIA, 28*(4), 890–894. doi:10.1093/jamia/ocaa268 PMID:33340404

Raj, R., & Kos, A. (2022). A Comprehensive Study of Mobile Robot: History, Developments, Applications, and Future Research Perspectives. *Applied Sciences (Basel, Switzerland), 12*(14), 6951. doi:10.3390/app12146951

Ramalingam, B., Yin, J., Rajesh Elara, M., Tamilselvam, Y. K., Mohan Rayguru, M., Muthugala, M. A. V. J., & Félix Gómez, B. (2020). A human support robot for the cleaning and maintenance of door handles using a deep-learning framework. *Sensors (Basel), 20*(12), 20. doi:10.3390/s20123543 PMID:32585864

Rasekhipour, Y., Khajepour, A., Chen, S.-K., & Litkouhi, B. (2017). A Potential Field-Based Model Predictive Path-Planning Controller for Autonomous Road Vehicles. *IEEE Transactions on Intelligent Transportation Systems, 2017*(18), 1255–1267. doi:10.1109/TITS.2016.2604240

Russell, S. (2022). *Artificial Intelligence: A Modern Approach* (3rd ed.). Pearson. Available online: https://zoo.cs.yale.edu/classes/cs470/materials/aima2010.pdf

Saike, J. (2022). Autonomous Navigation System of Greenhouse Mobile Robot Based on 3D Lidar and 2D Lidar SLAM. *Front. Plant Sci., 2022*(13), 815218. PMID:35360319

Sendak, M. P. (2020). A path for translation of machine learning products into healthcare delivery. *EMJ Innov., 10*, 19–00172.

Stojkoska, B. L. (2017). A review of Internet of Things for smart home: Challenges and solutions. *Journal of Cleaner Production, 140*, 1454-1464.

Tong, Y., Liu, H., & Zhang, Z. (2024, February). Advancements in humanoid robots: A comprehensive review and future prospects. *IEEE/CAA J. Autom. Sinica, 11*(2), 301–328. doi:10.1109/JAS.2023.124140

Topol, E. J. (2019). High-performance medicine: The convergence of human and artificial intelligence. *Nature Medicine, 25*(1), 44–56. doi:10.1038/s41591-018-0300-7 PMID:30617339

Vrontis, D., Christofi, M., Pereira, V., Tarba, S., Makrides, A., & Trichina, E. (2022). Artificial intelligence, robotics, advanced technologies and human resource management: A systematic review. *International Journal of Human Resource Management*, *33*(6), 1237–1266. Advance online publication. doi:10.1080/09585 192.2020.1871398

Wada, K. (2003). Effects of robot assisted activity to elderly people who stay at a health service facility for the aged. *IEEE Int Conf Intell Robots Syst.*, *3*, 2847–2852.

Xie, D., Chen, L., Liu, L., Chen, L., & Wang, H. (2022). Actuators and Sensors for Application in Agricultural Robots: A Review. *Machines*, *2022*(10), 913. doi:10.3390/ machines10100913

Xiong. (2024). Intuitive Human-Robot-Environment Interaction With EMG Signals: A Review. *IEEE/CAA Journal of Automatica Sinica*. . doi:10.1109/JAS.2024.124329

Yunjun Zheng. (2024). Adaptive Trajectory Tracking Control for Nonholonomic Wheeled Mobile Robots: A Barrier Function Sliding Mode Approach. *IEEE/CAA Journal of Automatica Sinica, 11*(4), 1007-1021. . doi:10.1109/JAS.2023.124002

Yurtsever, E. (2020). A Survey of Autonomous Driving: Common Practices and Emerging Technologies. *IEEE Access, 8*, 58443–58469.

Zaidan, A. A., Zaidan, B. B., Qahtan, M. Y., Albahri, O. S., Albahri, A. S., Alaa, M., Jumaah, F. M., Talal, M., Tan, K. L., Shir, W. L., & Lim, C. K. (2018). A survey on communication components for IoT-based technologies in smart homes. *Telecommunication Systems*, *69*(1), 1–25. doi:10.1007/s11235-018-0430-8

Conclusion

This chapter presents conclusions, while reflecting on 19th and 20th Centuries Industrial Revolution, the 21st Century Communication Technology and Beginning of Internet, Rise of New Digital Paradigm and Future Web and final Open Research Questions.

REFLECTION ON RISE FOR THE 19TH AND 20TH CENTURIES INDUSTRIAL REVOLUTION

Early years of 19th century were a new beginning of the Industrial Revolution side-by-side with invention including transportation technologies (i.e., Locomotive, Car, etc.), Electricity, Morse-Code, Telephone, plus. Figures 1 and 2, show the landscape of the Factory Age and Industrial Revolution, followed by Rise of AI, Ultra-Smart Humanoid Robotics and Cyberspace

Figure 1. The Factory Age (Google images)

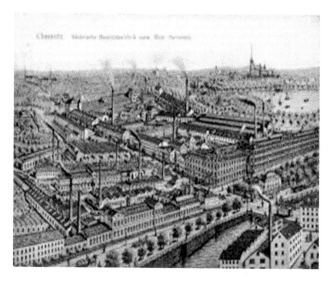

Figure 2. Industrial Revolution Overview (google images)

REFLECTION ON RISE FOR THE 21TH CENTURY COMMUNICATION TECHNOLOGIES AND BEGINNING OF INTERNET

The 19th, 20th and 21st Centuries were driven by scientific discoveries and applied research in the field of:

- Physics & Math
- Electricity
- Mores Code
- Telephone
- Locomotive - Car Industry
- World's Economic Crisis
- Electronics & Transistor
- Telephone
- Radio
- TVs
- Electric Drives
- Controlled Systems
- Microprocessor
- PC
- Computer Aided Engineering
- Internet
- Web

Apart from above listed technological advancements, past 21st century evolution in communications technologies brought a Wireless Communication Revolution shown in Figure which has become a backbone of todays' and tomorrows' Internet.

Figure 3. 21st Century Wireless Revolution
Credit: Dr. Rick Wietfeldt Texas Instruments

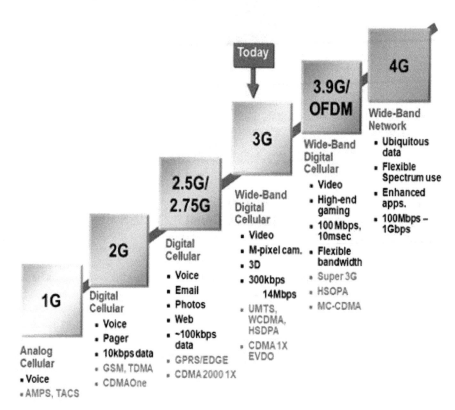

Kevin Kelly in "New Rules for the New Economy", stated: "Because communications – which in the end is what the digital technology and media are all about – is not just a sector of the economy. Communications is the economy. The new economy is about communications, deep and wide". Rise of ubiquitous Internet created a platform for New Digital Paradigm and Future IT Infrastructure shown in Figure 3.

REFLECTION ON RISE OF NEW DIGITAL PARADIGM AND FUTURE WEB

Figure 4. New Digital Paradigm
Credit: Korea Telecom

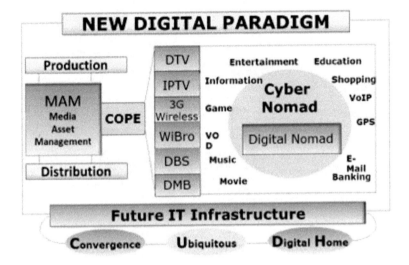

Figure 5. Future Web
Credit to Jeff Jaffe

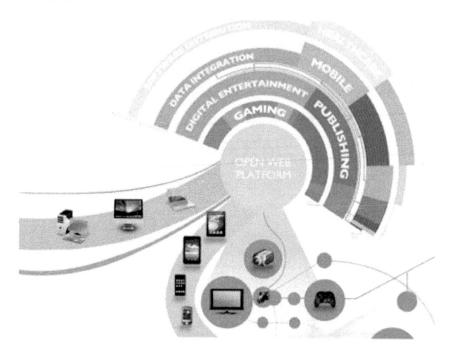

The end of 21st and current 22nd Centuries introduced three major technological areas

Smart: Intelligent Devices and Sensor Networks
Analytics: Distributed Intelligence
Humanoid Computing: Human to Computer Interaction and Integration

The Research, Innovation and Development were driven by:
Intelligent Device Integration: As devices gain in diversity, density and intelligence, so does the opportunity to gather knowledge.

Analytics and Insight: Exploiting emerging data sources for high performance.
Human Computer Interaction: The impact of emerging technologies and new business needs on workforce productivity and business performance.
Systems Integration: Exploring tomorrow's enterprise ICT systems.

Computing Industry trends were driven

Miniaturization: Everything has become smaller. ENIAC's old-fashioned radio-style vacuum tubes gave way after 1947 to the smaller, faster, more reliable transistor.

236

Speed: Due to enormous large volume of transaction and information processes the highest processing and communication speed is essential in all sectors.

Affordability: The cost is critical to all business worldwide.

REFLECTION ON RISE OF AI, ULTRA-SMART HUMANOID ROBOTICS AND CYBERSPACE

Given the current challenges in Industry, Business, Agriculture, Food production and distribution, National Security, Pollution, and Green House Effect in Slovakia and European Community, the practical applications of Artificial intelligence (AI) has become essential and well accepted in many sectors. The AI contributes side by side with Big Data facilitates data collection, processing, effective planning, decision-making, process optimization, and smart collaborative ecosystems.

The current wide range of applications of AI in conjunction with the Humanoid Robotics has already triggered lot of discussions about its potential impact on society, economy, business, industry, politics, education and other areas. Many would agree that applications of AI are very valuable in combating humanity current challenges such as the climate change, pollution, green-house effect, transportation, education, health, cyber security, national security, and much more.

However, some view may appear to be rather pessimistic which could be result of lack of knowledge of the AI fundamental functions and computational mechanisms.

The application of AI and Humanoid Robotics will have significant impact on the scientific communities in multiple areas of knowledge including, smart computing, applied engineering, mathematics, humanities and social studies. The Ultra-Smart Fully Automated Cyberspace will provide a platform for future research, innovation and development in the field of Ultra-Smart Computational Devices for the betterment of mankind.

FUTURE RESEARCH DIRECTIONS & OPEN QUESTIONS

In my view we may be still quite far from creating an Ultra-Smart Humanoid Robots, but very close to be able to imagine these Human-like Machines, and yet still asking following questions:

- Will the future AI based Humanoid Robotics reach the level of perfection that may much the human-like emotions, sense of values, touch, sensitivity, reasoning, trust and reliability?
- Why does the Humanoid Robot differ from us Humans?

- What will it take to make a Humanoid Robot that will be capable of becoming a fully accepted member of family and society?
- With rise of AI from Artificial General Intelligence (AGI) to future Artificial Super Intelligence (ASI), will the next Generation Ultra-Smart Robots contribute to betterment of mankind?

The book contributes to creation of new opportunities for Applied AI and Humanoid Robotics in the government, business, industry and academia both at local and national levels, as the EU Level. The book promotes further research, innovation and development in the field of Applied AI and Humanoid Robotics. The Applied AI with Humanoid Robotics may open a new Horizon for more effective Smart Governance, Business, Health, Education, Manufacturing Section, Cyber Security and National Security.

The book purpose is to create new opportunities for high-tech-start-up companies in the region. The project will promote high-tech innovation and the Applied AI standardization to shaping current policies concerning the AI related Ethical Norms. Hopefully, the book promotes better clarity and understanding of the Applied AI to community of Experts and general public. Hopefully book will promotes creation of Creation a research center for evaluating the impact of applied artificial intelligence.

Compilation of References

Abdollahi, H., Mahoor, M., Zandie, R., Sewierski, J., & Qualls, S. (2022). Artificial emotional intelligence in socially assistive robots for older adults: A pilot study. *IEEE Transactions on Affective Computing*. PMID:37840968

Admoni, H., & Scassellati, B. (2016). Social eye gaze in human-robot interaction: A review. *Journal of Human-Robot Interaction*, 6(1), 25–63. doi:10.5898/JHRI.6.1.Admoni

Agarwala, N. (2023). Robots and artificial intelligence in the military. *Obrana a Strategie (Defence and Strategy)*, 23(2), 83–100. doi:10.3849/1802-7199.23.2023.02.083-100

Agerskov, S. (2023). *Asger Balle Pedersen, Roman Beck, Ethical guidelines for Blockchain system*. ECIS.

Agrawal, A. V., Pitchai, R., Senthamaraikannan, C., Balaji, N. A., Sajithra, S., & Boopathi, S. (2023). Digital Education System During the COVID-19 Pandemic. In Using Assistive Technology for Inclusive Learning in K-12 Classrooms (pp. 104–126). IGI Global. doi:10.4018/978-1-6684-6424-3.ch005

Agrawal, A. V., Shashibhushan, G., Pradeep, S., Padhi, S., Sugumar, D., & Boopathi, S. (2023). Synergizing Artificial Intelligence, 5G, and Cloud Computing for Efficient Energy Conversion Using Agricultural Waste. In Sustainable Science and Intelligent Technologies for Societal Development (pp. 475–497). IGI Global.

Ahmad, J. (n.d.). *Navigating the Future: Harnessing Artificial Intelligence for Business Success*. Academic Press.

Ahmed, M. M., El-Sayed Seleman, M. M., Fydrych, D., & Çam, G. (2023). Friction stir welding of aluminum in the aerospace industry: The current progress and state-of-the-art review. *Materials (Basel)*, 16(8), 2971. doi:10.3390/ma16082971 PMID:37109809

Ahn, H. S. (2019). Hospital receptionist robot v2: design for enhancing verbal interaction with social skills. *2019 28th IEEE International Conference on Robot and Human Interactive Communication (RO-MAN)*, 1–6. 10.1109/RO-MAN46459.2019.8956300

Akinlabi, E. T., Mahamood, R. M., Akinlabi, E. T., & Mahamood, R. M. (2020). Future Research Direction in Friction Welding, Friction Stir Welding and Friction Stir Processing. *Solid-State Welding: Friction and Friction Stir Welding Processes*, 131–142.

Alatise, M.B. (2020). A Review on Challenges of Autonomous Mobile Robot and Sensor Fusion Methods. *IEEE Access, 8*, 39830–39846.

Aldoseri, A., Al-Khalifa, K., & Hamouda, A. (2023). *A roadmap for integrating automation with process optimization for AI-powered digital transformation.* Academic Press.

Aleksander, I. (2017). Partners of humans: A realistic assessment of the role of robots in the foreseeable future. *Journal of Information Technology, 32*(1), 1–9. doi:10.1057/s41265-016-0032-4

Alghowinem, S. (2021). Beyond the words: analysis and detection of self-disclosure behavior during robot positive psychology interaction. *2021 16th IEEE International Conference on Automatic Face and Gesture Recognition (FG 2021)*, 1–8. 10.1109/FG52635.2021.9666969

Ali, M. N., Senthil, T., Ilakkiya, T., Hasan, D. S., Ganapathy, N. B. S., & Boopathi, S. (2024). IoT's Role in Smart Manufacturing Transformation for Enhanced Household Product Quality. In *Advanced Applications in Osmotic Computing* (pp. 252–289). IGI Global. doi:10.4018/979-8-3693-1694-8.ch014

Allam Hamdan. (2021). The Relationship Between Intellectual Capital in the Fourth Industrial Revolution and Firm Performance in Jordan. *Studies in Computational Intelligence, 935*. https://www.springer.com/series/7092

Aly, A. (2012). Prosody-driven robot arm gestures generation in human-robot interaction. *Proceedings of the seventh annual ACM/IEEE international conference on Human-Robot Interaction*, 257–258. 10.1145/2157689.2157783

Anderson, T. L. (2017). *Fracture Mechanics—Fundamentals and Applications.* CRC. doi:10.1201/9781315370293

Andhare, P. (2016). Pick and place industrial robot controller with computer vision. *2016 International Conference on Computing Communication Control and automation (ICCUBEA)*, 1–4. 10.1109/ICCUBEA.2016.7860048

Anuradha, J. (2023). *Used of Artificial Intelligence & Robotics in Military Field.* Academic Press.

Arunprasad, R., Surendhiran, G., Ragul, M., Soundarrajan, T., Moutheepan, S., & Boopathi, S. (2018). Review on friction stir welding process. *International Journal of Applied Engineering Research: IJAER, 13*, 5750–5758.

Ashok, K. (2022). Collaborative analysis of audio-visual speech synthesis with sensor measurements for regulating human-robot interaction. *Int. J. Syst. Assur. Eng. Manag.*, 1–8. . doi:10.1007/s13198-022-01709-y

Aydin, Y., Tokatli, O., Patoglu, V., & Basdogan, C. (2020). A computational multicriteria optimization approach to controller design for physical human-robot interaction. *IEEE Transactions on Robotics*, *36*(6), 1791–1804. doi:10.1109/TRO.2020.2998606

Babulak, E. (2017). Cyber Security Solutions and Challenges in Ultrafast Internet Connecting Ultra-Smart Computational Devices. *ICSES Transactions on Computer Networks and Communications*, *3*(1).

Babulak, E. (2021). Third Millennium Life Saving Smart Cyberspace Driven by AI & Robotics. *COJ Rob Artificial Intel, 1*(4).

BabulakE. (2018). Role of Computational Physics in the Age Third Millennium Automation. *Available at*SSRN3404595.

Babulak, E. (2019). Third Millennium Smart Cyberspace. *ICSES*, *3*, 1–2.

Bagloee. (2016). Autonomous vehicles: challenges, opportunities, and future implications for transportation policies. *Journal of Modern Transportation, 24*, 284-303.

Bajwa, J., Munir, U., Nori, A., & Williams, B. (2021). Artificial intelligence in healthcare: Transforming the practice of medicine. *Future Healthcare Journal*, *8*(2), 188–194. doi:10.7861/fhj.2021-0095 PMID:34286183

Bal, Afacan, Clardy, & Cakir. (2023). Inclusive Future Making: Building a Culturally Responsive Behavioral Support System at an Urban Middle School with Local Stakeholders. Academic Press.

Bayar, A. (2023). Edge Computing Applications in Industrial IoT: A Literature Review. In J. Á. Bañares, J. Altmann, O. Agmon Ben-Yehuda, K. Djemame, V. Stankovski, & B. Tuffin (Eds.), Lecture Notes in Computer Science: Vol. 13430. *Economics of Grids, Clouds, Systems, and Services. GECON 2022.* Springer. doi:10.1007/978-3-031-29315-3_11

Benbya, H. (2020) Artificial Intelligence in Organizations: Current State and Future Opportunities. *MIS Quarterly Executive, 19*(4). doi:10.2139/ssrn.3741983

Benston, S. (2022). Walking a fine germline: Synthesizing public opinion and legal precedent to develop policy recommendations for heritable gene-editing. *Journal of Bioethical Inquiry*, *19*(3), 421–431. Advance online publication. doi:10.1007/s11673-022-10186-8 PMID:35438443

Bera, K., Schalper, K. A., Rimm, D. L., Velcheti, V., & Madabhushi, A. (2019). Artificial intelligence in digital pathology - new tools for diagnosis and precision oncology. *Nature Reviews. Clinical Oncology*, *16*(11), 703–715. doi:10.1038/s41571-019-0252-y PMID:31399699

Bharadiya, J. P. (2023). Machine learning and AI in business intelligence: Trends and opportunities. [IJC]. *International Journal of Computer*, *48*(1), 123–134.

Bhushan, T. (2024). Artificial Intelligence, Cyberspace and International Law. *Indonesian Journal of International Law*, *21*(2), 3. https://scholarhub.ui.ac.id/ijil/vol21/iss2/3

Bigg, G., & Billings, S. (2014). The iceberg risk in the *Titanic* year of 1912: Was it exceptional? *Significance, 11*(3), 6–10. doi:10.1111/j.1740-9713.2014.00746.x

Boopathi, S. (2023). Deep Learning Techniques Applied for Automatic Sentence Generation. In Promoting Diversity, Equity, and Inclusion in Language Learning Environments (pp. 255–273). IGI Global. doi:10.4018/978-1-6684-3632-5.ch016

Boopathi, S. (2023). Securing Healthcare Systems Integrated With IoT: Fundamentals, Applications, and Future Trends. In Dynamics of Swarm Intelligence Health Analysis for the Next Generation (pp. 186–209). IGI Global.

Boopathi, S. (2024). Sustainable Development Using IoT and AI Techniques for Water Utilization in Agriculture. In Sustainable Development in AI, Blockchain, and E-Governance Applications (pp. 204–228). IGI Global. doi:10.4018/979-8-3693-1722-8.ch012

Boopathi, S., Kumar, P. K. S., Meena, R. S., Sudhakar, M., & Associates. (2023). Sustainable Developments of Modern Soil-Less Agro-Cultivation Systems: Aquaponic Culture. In Human Agro-Energy Optimization for Business and Industry (pp. 69–87). IGI Global.

Boopathi, S. (2024). Advancements in Machine Learning and AI for Intelligent Systems in Drone Applications for Smart City Developments. In *Futuristic e-Governance Security With Deep Learning Applications* (pp. 15–45). IGI Global. doi:10.4018/978-1-6684-9596-4.ch002

Boopathi, S., & Davim, J. P. (2023). *Sustainable Utilization of Nanoparticles and Nanofluids in Engineering Applications*. IGI Global. doi:10.4018/978-1-6684-9135-5

Boopathi, S., & Kanike, U. K. (2023). Applications of Artificial Intelligent and Machine Learning Techniques in Image Processing. In *Handbook of Research on Thrust Technologies' Effect on Image Processing* (pp. 151–173). IGI Global. doi:10.4018/978-1-6684-8618-4.ch010

Boopathi, S., & Khang, A. (2023). AI-Integrated Technology for a Secure and Ethical Healthcare Ecosystem. In *AI and IoT-Based Technologies for Precision Medicine* (pp. 36–59). IGI Global. doi:10.4018/979-8-3693-0876-9.ch003

Boopathi, S., Kumaresan, A., & Manohar, N., & KrishnaMoorthi, R. (2017). Review on Effect of Process Parameters—Friction Stir Welding Process. *International Research Journal of Engineering and Technology, 4*(7), 272–278.

Booth, K., Subramanyan, H., Liu, J., & Lukic, S. M. (2020). Parallel frameworks for robust optimization of medium-frequency transformers. *IEEE Journal of Emerging and Selected Topics in Power Electronics, 9*(4), 5097–5112. doi:10.1109/JESTPE.2020.3042527

Börner, K., Scrivner, O., Cross, L. E., Gallant, M., Ma, S., Martin, A. S., Record, L., Yang, H., & Dilger, J. M. (2020). Mapping the co-evolution of artificial intelligence, robotics, and the internet of things over 20 years (1998-2017). *PLoS One, 15*(12), e0242984. doi:10.1371/journal.pone.0242984 PMID:33264328

Branco, R., Prates, P., Costa, J. D. M., Berto, F., & Kotousov, A. (2018). New methodology of fatigue life evaluation for multiaxially loaded notched components based on two uniaxial strain-controlled tests. *International Journal of Fatigue, 111*, 308–320. doi:10.1016/j.ijfatigue.2018.02.027

Campbell, F. C. (Ed.), *Elements of Metallurgy and Engineering Alloys.* doi:10.31399/asm.tb.emea.9781627082518

Carlos Flavián. (2019). The impact of virtual, augmented and mixed reality technologies on the customer experience. *Journal of Business Research, 100*, 547-560. doi:10.1016/j.jbusres.2018.10.050

Cengel, Y., Boles, M., & Kanoglu, M. (2018). *Thermodynamics: An Engineering Approach* (9th ed.). McGraw—Hill.

Chevalier, P., Kompatsiari, K., Ciardo, F., & Wykowska, A. (2020). Examining joint attention with the use of humanoid robots-A new approach to study fundamental mechanisms of social cognition. *Psychonomic Bulletin & Review, 27*(2), 217–236. doi:10.3758/s13423-019-01689-4 PMID:31848909

Czeczot, G., Rojek, I., Mikołajewski, D., & Sangho, B. (2023). AI in IIoT Management of Cybersecurity for Industry 4.0 and Industry 5.0 Purposes. *Electronics (Basel), 2023*(12), 3800. doi:10.3390/electronics12183800

D'UrsoF. (2023) Revolutionizing medical education: the impact of Virtual and Mixed Reality on training and skill acquisition.Preprints2023, 2023121901. https://doi.org/ doi:10.20944/preprints202312.1901.v2

David, J. G. (2018). *Introduction to Quantum Mechanics.* Cambridge University Press.

Deo, N., & Anjankar, A. (2023). Artificial Intelligence with robotics in Healthcare: A narrative review of its viability in India. *Cureus.* Advance online publication. doi:10.7759/cureus.39416 PMID:37362504

Dhanalakshmi, M., Tamilarasi, K., Saravanan, S., Sujatha, G., Boopathi, S., & Associates. (2024). Fog Computing-Based Framework and Solutions for Intelligent Systems: Enabling Autonomy in Vehicles. In Computational Intelligence for Green Cloud Computing and Digital Waste Management (pp. 330–356). IGI Global.

Dian, M. G. (2015). The Next Level of Energy Efficiency: The Five Challenges Ahead. *The Electricity Journal, 28*(7), 44-56. doi:10.1016/j.tej.2015.07.001

Doncieux, S., Chatila, R., Straube, S., & Kirchner, F. (2022). Human-centered AI and robotics. *AI Perspect, 4*(1), 1. doi:10.1186/s42467-021-00014-x

Duga, J. J., Fisher, W. H., Buxaum, R. W., Rosenfield, A. R., Buhr, A. R., Honton, E. J., & McMillan, S. C. (1982). *The Economic Effects of Fracture in the United States.* Final Report, September 30. Battelle Laboratories, Columbus, OH. Available as NBS Special Publication 647-2.

Elsayed, E. A. (2012). *Reliability Engineering.* John Wiley & Sons.

Ethical, K. E. (2022, September). Legal and Social Implications of Emerging Technology (ELSIET) Symposium. *Journal of Bioethical Inquiry, 19*(3), 363–370. doi:10.1007/s11673-022-10197-5 PMID:35749026

Ferrera, M. (2018). The Aqualoc Dataset: Towards Real-Time Underwater Localization from a Visual-Inertial-Pressure Acquisition System. arXiv 2018, arXiv:1809.07076.

Garvin, D. A. (1987). Competing on the Eight Dimensions of Quality. *Harvard Business Review, 65*(6), 101–109.

Gasteiger, N., Hellou, M., & Ahn, H. S. (2023). Factors for personalization and localization to optimize human–robot interaction: A literature review. *International Journal of Social Robotics, 15*(4), 689–701. doi:10.1007/s12369-021-00811-8

Geng, R., Li, M., Hu, Z., Han, Z., & Zheng, R. (2022). Digital Twin in smart manufacturing: Remote control and virtual machining using VR and AR technologies. *Structural and Multidisciplinary Optimization, 65*(11), 321. doi:10.1007/s00158-022-03426-3

Ghayvat, H., Mukhopadhyay, S. C., Liu, J., Babu, A., Elahi, E., & Gui, X. (2015). Internet of Things for smart homes and buildings: Opportunities and Challenges. *Journal of Telecommunications and the Digital Economy, 3*(4), 33–47. doi:10.18080/jtde.v3n4.23

Gill, A., Khalid, A., & Murugan, R. (2023). A Crucial Review on Ai Use In Hybrid Grids: A Ultra Smart Way Towards Micro Grid. *2023 International Conference on Power Energy, Environment & Intelligent Control (PEEIC)*, 1364–1368. 10.1109/PEEIC59336.2023.10451299

González-García, Gómez-Espinosa, A., Cuan-Urquizo, E., García-Valdovinos, L. G., Salgado-Jiménez, T., & Cabello, J. A. E. (2020). Autonomous Underwater Vehicles: Localization, Navigation, and Communication for Collaborative Missions. *Applied Sciences (Basel, Switzerland), 2020*(10), 1256. doi:10.3390/app10041256

Goodno, B. J., & Gere, J. M. (2017). *Mechanics of Materials* (9th ed.). Cengage Learning, Inc.

Gracias, J. S., Parnell, G. S., Specking, E., Pohl, E. A., & Buchanan, R. (2023). Smart Cities-A Structured Literature Review. *Smart Cities, 2023*(6), 1719–1743. doi:10.3390/smartcities6040080

Grove, A. (1967). *Physics and Technology of Semiconductor Device* (International Edition). Wiley.

Gunasekaran, K., Boopathi, S., & Sureshkumar, M. (2022). Analysis of a Cryogenically Cooled Near-Dry WEDM Process using Different Dielectrics. *Mater-Tehnol.Si. Materials Technology, 56*(2), 179–186.

Hahn, G. J., & Meeker, W. Q. (2004). *How to Plan an Accelerated Life Test (E-Book)*. ASQ Quality Press.

Haque, A., Milstein, A., & Fei-Fei, L. (2020). Illuminating the dark spaces of healthcare with ambient intelligence. *Nature, 585*(7824), 193–202. doi:10.1038/s41586-020-2669-y PMID:32908264

He, A.-Z., & Zhang, Y. (2023). AI-powered touch points in the customer journey: A systematic literature review and research agenda. *Journal of Research in Interactive Marketing*, *17*(4), 620–639. doi:10.1108/JRIM-03-2022-0082

Hlee, S., Park, J., Park, H., Koo, C., & Chang, Y. (2023). Understanding customer's meaningful engagement with AI-powered service robots. *Information Technology & People*, *36*(3), 1020–1047. doi:10.1108/ITP-10-2020-0740

Hunde, B. R., & Woldeyohannes, A. D. (2022). Future prospects of computer-aided design (CAD)–A review from the perspective of artificial intelligence (AI), extended reality, and 3D printing. *Results in Engineering*, *14*, 100478. doi:10.1016/j.rineng.2022.100478

Hussain, Z., Babe, M., Saravanan, S., Srimathy, G., Roopa, H., & Boopathi, S. (2023). Optimizing Biomass-to-Biofuel Conversion: IoT and AI Integration for Enhanced Efficiency and Sustainability. In Circular Economy Implementation for Sustainability in the Built Environment (pp. 191–214). IGI Global.

IEEE Standard Glossary of Software Engineering Terminology. (2002). *IEEE STD 610.12-1990. Standards Coordinating Committee of the Computer Society of IEEE.* Available online: https://ieeexplore.ieee.org/document/159342

Jarrahi, M. H., Kenyon, S., Brown, A., Donahue, C., & Wicher, C. (2023). Artificial intelligence: A strategy to harness its power through organizational learning. *The Journal of Business Strategy*, *44*(3), 126–135. doi:10.1108/JBS-11-2021-0182

Kannadhasan, S., & Nagarajan, R. (2022). Recent trends in wearable technologies, challenges and opportunities. *Designing Intelligent Healthcare Systems, Products, and Services Using Disruptive Technologies and Health Informatics*, 131–143.

Karami, V., Yaffe, M. J., & Rahimi, S. A. (2023). Early Detection of Alzheimer's Disease Assisted by AI-Powered Human-Robot Communication. In *Machine Learning and Artificial Intelligence in Healthcare Systems* (pp. 331–348). CRC Press.

Karlsson, M., Bagge Carlson, F., Holmstrand, M., Robertsson, A., De Backer, J., Quintino, L., Assuncao, E., & Johansson, R. (2023). Robotic friction stir welding–seam-tracking control, force control and process supervision. *The Industrial Robot*, *50*(5), 722–730. doi:10.1108/IR-06-2022-0153

Karnopp, D. C., Margolis, D. L., & Rosenberg, R. C. (2012). *System Dynamics: Modeling, Simulation, and Control of Mechatronic Systems*. John Wiley & Sons. doi:10.1002/9781118152812

KAV, R. P., Pandraju, T. K. S., Boopathi, S., Saravanan, P., Rathan, S. K., & Sathish, T. (2023). Hybrid Deep Learning Technique for Optimal Wind Mill Speed Estimation. *2023 7th International Conference on Electronics, Communication and Aerospace Technology (ICECA)*, 181–186.

Khalid, U., Naeem, M., Stasolla, F., Syed, M., Abbas, M., & Coronato, A. (2024). Impact of AI-Powered Solutions in Rehabilitation Process: Recent Improvements and Future Trends. *International Journal of General Medicine*, *17*, 943–969. doi:10.2147/IJGM.S453903 PMID:38495919

Kiran Jot Singh. (2021). All about human-robot interaction. In Cognitive Data Science in Sustainable Computing, Cognitive Computing for Human-Robot Interaction. Academic Press. doi:10.1016/B978-0-323-85769-7.00010-0

Kitsios, F., & Kamariotou, M. (2021). Artificial intelligence and business strategy towards digital transformation: A research agenda. *Sustainability (Basel)*, *13*(4), 2025. doi:10.3390/su13042025

Koshariya, A. K., Kalaiyarasi, D., Jovith, A. A., Sivakami, T., Hasan, D. S., & Boopathi, S. (2023). AI-Enabled IoT and WSN-Integrated Smart Agriculture System. In *Artificial Intelligence Tools and Technologies for Smart Farming and Agriculture Practices* (pp. 200–218). IGI Global. doi:10.4018/978-1-6684-8516-3.ch011

Koshariya, A. K., Khatoon, S., Marathe, A. M., Suba, G. M., Baral, D., & Boopathi, S. (2023). Agricultural Waste Management Systems Using Artificial Intelligence Techniques. In *AI-Enabled Social Robotics in Human Care Services* (pp. 236–258). IGI Global. doi:10.4018/978-1-6684-8171-4.ch009

Kumar, M., Kumar, K., Sasikala, P., Sampath, B., Gopi, B., & Sundaram, S. (2023). Sustainable Green Energy Generation From Waste Water: IoT and ML Integration. In Sustainable Science and Intelligent Technologies for Societal Development (pp. 440–463). IGI Global.

Kumara, V., Sharma, M. D., Samson Isaac, J., Saravanan, S., Suganthi, D., & Boopathi, S. (2023). An AI-Integrated Green Power Monitoring System: Empowering Small and Medium Enterprises. In Advances in Environmental Engineering and Green Technologies (pp. 218–244). IGI Global. doi:10.4018/979-8-3693-0338-2.ch013

Kushwah, J. S., Gupta, M., Shrivastava, S., Saxena, N., Saini, R., & Boopathi, S. (2024). Psychological Impacts, Prevention Strategies, and Intervention Approaches Across Age Groups: Unmasking Cyberbullying. In Change Dynamics in Healthcare, Technological Innovations, and Complex Scenarios (pp. 89–109). IGI Global.

Lamontagne, L. (2014). A framework for building adaptive intelligent virtual assistants. *Artificial Intelligence and Applications (Commerce, Calif.)*, *10*.

Liu, X., He, X., Wang, M., & Shen, H. (2022). What influences patients' continuance intention to use AI-powered service robots at hospitals? The role of individual characteristics. *Technology in Society*, *70*, 101996. doi:10.1016/j.techsoc.2022.101996

Liu, Y., & Ren, L. (2023). *Artificial Intelligence Techniques for Joint Sensing and Localization in Future Wireless Networks*. Hindawi.

Lopes & Alexandre. (2023). An Overview of Blockchain Integration with Robotics and Artificial Intelligence. Cornell University.

Luo, H., Zhao, F., Guo, S., Yu, C., Liu, G., & Wu, T. (2021). Mechanical performance research of friction stir welding robot for aerospace applications. *International Journal of Advanced Robotic Systems*, *18*(1), 1729881421996543. doi:10.1177/1729881421996543

Magaziner, I. C., & Patinkin, M. (1989). Cold competition: GE wages the refrigerator war. *Harvard Business Review*, *89*(2), 114–124.

Maguluri, L. P., Arularasan, A., & Boopathi, S. (2023). Assessing Security Concerns for AI-Based Drones in Smart Cities. In Effective AI, Blockchain, and E-Governance Applications for Knowledge Discovery and Management (pp. 27–47). IGI Global. doi:10.4018/978-1-6684-9151-5.ch002

Maheswari, B. U., Imambi, S. S., Hasan, D., Meenakshi, S., Pratheep, V., & Boopathi, S. (2023). Internet of things and machine learning-integrated smart robotics. In Global Perspectives on Robotics and Autonomous Systems: Development and Applications (pp. 240–258). IGI Global. doi:10.4018/978-1-6684-7791-5.ch010

Malathi, J., Kusha, K., Isaac, S., Ramesh, A., Rajendiran, M., & Boopathi, S. (2024). IoT-Enabled Remote Patient Monitoring for Chronic Disease Management and Cost Savings: Transforming Healthcare. In Advances in Explainable AI Applications for Smart Cities (pp. 371–388). IGI Global.

Marcos-Pablos, S., & García-Peñalvo, F. J. (2022). Emotional intelligence in robotics: A scoping review. *New Trends in Disruptive Technologies. Tech Ethics and Artificial Intelligence: The DITTET Collection*, *1*, 66–75.

Martínez-Rojas, A., Sánchez-Oliva, J., López-Carnicer, J. M., & Jiménez-Ramírez, A. (2021). Airpa: An architecture to support the execution and maintenance of AI-powered RPA robots. *International Conference on Business Process Management*, 38–48. 10.1007/978-3-030-85867-4_4

McEnroe, P., Wang, S., & Liyanage, M. (2022). A survey on the convergence of edge computing and AI for UAVs: Opportunities and challenges. *IEEE Internet of Things Journal*, *9*(17), 15435–15459. doi:10.1109/JIOT.2022.3176400

McPherson, J. (1989) Accelerated Testing. In Electronic Materials Handbook; Volume 1: Packaging. ASM International Publishing.

Mendes, N., Neto, P., Loureiro, A., & Moreira, A. P. (2016). Machines and control systems for friction stir welding: A review. *Materials & Design*, *90*, 256–265. doi:10.1016/j.matdes.2015.10.124

Mendes, N., Neto, P., Simão, M., Loureiro, A., & Pires, J. (2016). A novel friction stir welding robotic platform: Welding polymeric materials. *International Journal of Advanced Manufacturing Technology*, *85*(1-4), 37–46. doi:10.1007/s00170-014-6024-z

Mishra, D., Roy, R. B., Dutta, S., Pal, S. K., & Chakravarty, D. (2018). A review on sensor based monitoring and control of friction stir welding process and a roadmap to Industry 4.0. *Journal of Manufacturing Processes*, *36*, 373–397. doi:10.1016/j.jmapro.2018.10.016

Mohanty, A., Venkateswaran, N., Ranjit, P., Tripathi, M. A., & Boopathi, S. (2023). Innovative Strategy for Profitable Automobile Industries: Working Capital Management. In Handbook of Research on Designing Sustainable Supply Chains to Achieve a Circular Economy (pp. 412–428). IGI Global.

Mohanty, A., Jothi, B., Jeyasudha, J., Ranjit, P., Isaac, J. S., & Boopathi, S. (2023). Additive Manufacturing Using Robotic Programming. In *AI-Enabled Social Robotics in Human Care Services* (pp. 259–282). IGI Global. doi:10.4018/978-1-6684-8171-4.ch010

Montgomery, D. (2013). *Design and Analysis of Experiments* (8th ed.). John Wiley and Son.

Moretti, C. B., Delbem, A. C., & Krebs, H. I. (2020). Human-robot interaction: Kinematic and kinetic data analysis framework. *2020 8th IEEE RAS/EMBS International Conference for Biomedical Robotics and Biomechatronics (BioRob)*, 235–239.

Mukherjee, D., Gupta, K., & Najjaran, H. (2022). An ai-powered hierarchical communication framework for robust human-robot collaboration in industrial settings. *2022 31st IEEE International Conference on Robot and Human Interactive Communication (RO-MAN)*, 1321–1326.

Naneva, S., Sarda Gou, M., Webb, T. L., & Prescott, T. J. (2020). A Systematic Review of Attitudes, Anxiety, Acceptance, and Trust towards social Robots. *International Journal of Social Robotics*, *2020*(12), 1179–1201. doi:10.1007/s12369-020-00659-4

Naveeenkumar, N., Rallapalli, S., Sasikala, K., Priya, P. V., Husain, J., & Boopathi, S. (2024). Enhancing Consumer Behavior and Experience Through AI-Driven Insights Optimization. In *AI Impacts in Digital Consumer Behavior* (pp. 1–35). IGI Global. doi:10.4018/979-8-3693-1918-5.ch001

Nguyen, T.-H., Tran, D.-N., Vo, D.-L., Mai, V.-H., & Dao, X.-Q. (2022). AI-powered university: Design and deployment of robot assistant for smart universities. *Journal of Advances in Information Technology, 13*(1).

Okagbue, E. F., Muhideen, S., Anulika, A. G., Nchekwubemchukwu, I. S., Chinemerem, O. G., Tsakuwa, M. B., Achaa, L. O., Adarkwah, M. A., Funmi, K. B., Nneoma, N. C., & Mwase, C. (2023). An in-depth analysis of humanoid robotics in Higher Education System. *Education and Information Technologies*, *29*(1), 185–217. doi:10.1007/s10639-023-12263-w

Okamoto, E. (2021). Overview of nonlinear signal processing in 5G and 6G access technologies. *Nonlinear Theory and Its Applications, IEICE, 12*(3), 257–274. doi:10.1587/nolta.12.257

Ozturkcan, S., & Merdin-Uygur, E. (2021). Humanoid Service Robots: The future of healthcare? *Journal of Information Technology Teaching Cases, 12*(2), 163–169. doi:10.1177/20438869211003905

Pachiappan, K., Anitha, K., Pitchai, R., Sangeetha, S., Satyanarayana, T., & Boopathi, S. (2024). Intelligent Machines, IoT, and AI in Revolutionizing Agriculture for Water Processing. In *Handbook of Research on AI and ML for Intelligent Machines and Systems* (pp. 374–399). IGI Global.

Palmgren, A. G. (1924). Die Lebensdauer von Kugellagern. *Z. Ver. Dtsch. Ing., 68*, 339–341.

Patil, S., & Shankar, H. (2023). Transforming healthcare: Harnessing the power of AI in the modern era. *International Journal of Multidisciplinary Sciences and Arts, 2*(1), 60–70.

Pepito, J. A., Ito, H., Betriana, F., Tanioka, T., & Locsin, R. C. (2020). Intelligent humanoid robots expressing artificial humanlike empathy in nursing situations. *Nursing Philosophy*, *21*(4), e12318. doi:10.1111/nup.12318 PMID:33462939

Pérez, L., Rodríguez-Jiménez, S., Rodríguez, N., Usamentiaga, R., García, D. F., & Wang, L. (2019). Symbiotic human–robot collaborative approach for increased productivity and enhanced safety in the aerospace manufacturing industry. *International Journal of Advanced Manufacturing Technology*, *106*(3–4), 851–863. doi:10.1007/s00170-019-04638-6

Pitchai, R., Guru, K. V., Gandhi, J. N., Komala, C. R., Kumar, J. R. D., & Boopathi, S. (2024). Fog Computing-Integrated ML-Based Framework and Solutions for Intelligent Systems: Digital Healthcare Applications. In Technological Advancements in Data Processing for Next Generation Intelligent Systems (pp. 196–224). IGI Global. doi:10.4018/979-8-3693-0968-1.ch008

Potter, Oluwaseyi, & Olaoye. (2024). Advances in Artificial Intelligence for Robotics. *Journal of Robotic Systems*.

Prabhakar, D., Korgal, A., Shettigar, A. K., Herbert, M. A., Chandrashekharappa, M. P. G., Pimenov, D. Y., & Giasin, K. (2023). A Review of Optimization and Measurement Techniques of the Friction Stir Welding (FSW) Process. *Journal of Manufacturing and Materials Processing*, *7*(5), 181. doi:10.3390/jmmp7050181

Prabhuswamy, M., Tripathi, R., Vijayakumar, M., Thulasimani, T., Sundharesalingam, P., & Sampath, B. (2024). A Study on the Complex Nature of Higher Education Leadership: An Innovative Approach. In *Challenges of Globalization and Inclusivity in Academic Research* (pp. 202–223). IGI Global. doi:10.4018/979-8-3693-1371-8.ch013

Pradhan, B., Bharti, D., Chakravarty, S., Ray, S. S., Voinova, V. V., Bonartsev, A. P., & Pal, K. (2021). Internet of things and robotics in transforming current-day healthcare services. *Journal of Healthcare Engineering*, *2021*, 9999504. doi:10.1155/2021/9999504 PMID:34104368

Prakash, N. (2023). Merging Minds and Machines: The Role of Advancing AI in Robotics. EAI Endorsed Trans IoT, 10.

Pramila, P., Amudha, S., Saravanan, T., Sankar, S. R., Poongothai, E., & Boopathi, S. (2023). Design and Development of Robots for Medical Assistance: An Architectural Approach. In Contemporary Applications of Data Fusion for Advanced Healthcare Informatics (pp. 260–282). IGI Global.

PricewaterhouseCoopers. (n.d.). *No longer science fiction, AI and Robotics Are Transforming Healthcare*. PwC. https://www.pwc.com/gx/en/industries/healthcare/publications/ai-robotics-new-health/transforming-healthcare.html

Puranik, T. A., Shaik, N., Vankudoth, R., Kolhe, M. R., Yadav, N., & Boopathi, S. (2024). Study on Harmonizing Human-Robot (Drone) Collaboration: Navigating Seamless Interactions in Collaborative Environments. In Cybersecurity Issues and Challenges in the Drone Industry (pp. 1–26). IGI Global.

Quinn, T. P., Senadeera, M., Jacobs, S., Coghlan, S., & Le, V. (2021). Trust and medical AI: The challenges we face and the expertise needed to overcome them. *Journal of the American Medical Informatics Association : JAMIA*, *28*(4), 890–894. doi:10.1093/jamia/ocaa268 PMID:33340404

Rahamathunnisa, U., Sudhakar, K., Padhi, S., Bhattacharya, S., Shashibhushan, G., & Boopathi, S. (2024). Sustainable Energy Generation From Waste Water: IoT Integrated Technologies. In Adoption and Use of Technology Tools and Services by Economically Disadvantaged Communities: Implications for Growth and Sustainability (pp. 225–256). IGI Global.

Rahamathunnisa, U., Sudhakar, K., Murugan, T. K., Thivaharan, S., Rajkumar, M., & Boopathi, S. (2023). Cloud Computing Principles for Optimizing Robot Task Offloading Processes. In *AI-Enabled Social Robotics in Human Care Services* (pp. 188–211). IGI Global. doi:10.4018/978-1-6684-8171-4.ch007

Raj, A., Chadha, U., Chadha, A., Mahadevan, R. R., Sai, B. R., Chaudhary, D., Selvaraj, S. K., Lokeshkumar, R., Das, S., Karthikeyan, B., & others. (2023a). Weld quality monitoring via machine learning-enabled approaches. *International Journal on Interactive Design and Manufacturing*, 1–43.

Rajagopal, M., & Babu, M.N. (2018). *Virtual Teaching Assistant to Support Students' Efforts in Programming*. Academic Press.

Raj, R., & Kos, A. (2022). A Comprehensive Study of Mobile Robot: History, Developments, Applications, and Future Research Perspectives. *Applied Sciences (Basel, Switzerland)*, *12*(14), 6951. doi:10.3390/app12146951

Ramalingam, B., Yin, J., Rajesh Elara, M., Tamilselvam, Y. K., Mohan Rayguru, M., Muthugala, M. A. V. J., & Félix Gómez, B. (2020). A human support robot for the cleaning and maintenance of door handles using a deep-learning framework. *Sensors (Basel)*, *20*(12), 20. doi:10.3390/s20123543 PMID:32585864

Ramudu, K., Mohan, V. M., Jyothirmai, D., Prasad, D., Agrawal, R., & Boopathi, S. (2023). Machine Learning and Artificial Intelligence in Disease Prediction: Applications, Challenges, Limitations, Case Studies, and Future Directions. In Contemporary Applications of Data Fusion for Advanced Healthcare Informatics (pp. 297–318). IGI Global.

Rasekhipour, Y., Khajepour, A., Chen, S.-K., & Litkouhi, B. (2017). A Potential Field-Based Model Predictive Path-Planning Controller for Autonomous Road Vehicles. *IEEE Transactions on Intelligent Transportation Systems*, *2017*(18), 1255–1267. doi:10.1109/TITS.2016.2604240

Ravisankar, A., Sampath, B., & Asif, M. M. (2023). Economic Studies on Automobile Management: Working Capital and Investment Analysis. In Multidisciplinary Approaches to Organizational Governance During Health Crises (pp. 169–198). IGI Global.

Ravisankar, A., Shanthi, A., Lavanya, S., Ramaratnam, M., Krishnamoorthy, V., & Boopathi, S. (2024). Harnessing 6G for Consumer-Centric Business Strategies Across Electronic Industries. In AI Impacts in Digital Consumer Behavior (pp. 241–270). IGI Global.

Rebecca, B., Kumar, K. P. M., Padmini, S., Srivastava, B. K., Halder, S., & Boopathi, S. (2024). Convergence of Data Science-AI-Green Chemistry-Affordable Medicine: Transforming Drug Discovery. In *Handbook of Research on AI and ML for Intelligent Machines and Systems* (pp. 348–373). IGI Global.

Reddy, M. A., Gaurav, A., Ushasukhanya, S., Rao, V. C. S., Bhattacharya, S., & Boopathi, S. (2023). Bio-Medical Wastes Handling Strategies During the COVID-19 Pandemic. In *Multidisciplinary Approaches to Organizational Governance During Health Crises* (pp. 90–111). IGI Global. doi:10.4018/978-1-7998-9213-7.ch006

Reddy, S. (2023, October 26). *NLP in Robotics: Enhancing Human-Robot Interaction*. Pss Blog. https://www.pranathiss.com/blog/nlp-human-robot-interaction/

Reddy, J. N. (2021). *An Introduction to Nonlinear Finite Element Method with Applications to Heat Transfer, Fluid Mechanics, and Solid Mechanics*. Oxford Press.

Revathi, S., Babu, M., Rajkumar, N., Meti, V. K. V., Kandavalli, S. R., & Boopathi, S. (2024). Unleashing the Future Potential of 4D Printing: Exploring Applications in Wearable Technology, Robotics, Energy, Transportation, and Fashion. In *Human-Centered Approaches in Industry 5.0: Human-Machine Interaction, Virtual Reality Training, and Customer Sentiment Analysis* (pp. 131–153). IGI Global.

Rossos, D., Mihailidis, A., & Laschowski, B. (2023). AI-powered smart glasses for sensing and recognition of human-robot walking environments. bioRxiv, 2023–10. doi:10.1101/2023.10.24.563804

Russell, S. (2022). *Artificial Intelligence: A Modern Approach* (3rd ed.). Pearson. Available online: https://zoo.cs.yale.edu/classes/cs470/materials/aima2010.pdf

Saike, J. (2022). Autonomous Navigation System of Greenhouse Mobile Robot Based on 3D Lidar and 2D Lidar SLAM. *Front. Plant Sci.*, 2022(13), 815218. PMID:35360319

Samikannu, R., Koshariya, A. K., Poornima, E., Ramesh, S., Kumar, A., & Boopathi, S. (2022). Sustainable Development in Modern Aquaponics Cultivation Systems Using IoT Technologies. In *Human Agro-Energy Optimization for Business and Industry* (pp. 105–127). IGI Global.

Sampath, B., & Haribalaji, V. (2021). Influences of Welding Parameters on Friction Stir Welding of Aluminum and Magnesium: A Review. *Materials Research Proceedings*, 19(1), 322–330.

Sangeetha, M., Kannan, S. R., Boopathi, S., Ramya, J., Ishrat, M., & Sabarinathan, G. (2023). Prediction of Fruit Texture Features Using Deep Learning Techniques. *2023 4th International Conference on Smart Electronics and Communication (ICOSEC)*, 762–768.

Satav, S. D., Hasan, D. S., Pitchai, R., Mohanaprakash, T., Sultanuddin, S., & Boopathi, S. (2023). Next generation of internet of things (ngiot) in healthcare systems. In *Sustainable Science and Intelligent Technologies for Societal Development* (pp. 307–330). IGI Global.

Schaefer, S., Leung, K., Ivanovic, B., & Pavone, M. (2021). Leveraging neural network gradients within trajectory optimization for proactive human-robot interactions. *2021 IEEE International Conference on Robotics and Automation (ICRA)*, 9673–9679. 10.1109/ICRA48506.2021.9561443

Sendak, M. P. (2020). A path for translation of machine learning products into healthcare delivery. *EMJ Innov.*, *10*, 19–00172.

Senthil, T., Puviyarasan, M., Babu, S. R., Surakasi, R., Sampath, B., & Associates. (2023). Industrial Robot-Integrated Fused Deposition Modelling for the 3D Printing Process. In Development, Properties, and Industrial Applications of 3D Printed Polymer Composites (pp. 188–210). IGI Global.

Shaikh, T. A., Rasool, T., & Lone, F. R. (2022). Towards leveraging the role of machine learning and artificial intelligence in precision agriculture and smart farming. *Computers and Electronics in Agriculture*, *198*, 107119. doi:10.1016/j.compag.2022.107119

Sharma, D. M., Ramana, K. V., Jothilakshmi, R., Verma, R., Maheswari, B. U., & Boopathi, S. (2024). Integrating Generative AI Into K-12 Curriculums and Pedagogies in India: Opportunities and Challenges. *Facilitating Global Collaboration and Knowledge Sharing in Higher Education With Generative AI*, 133–161.

Sharma, M., Sharma, M., Sharma, N., & Boopathi, S. (2024). Building Sustainable Smart Cities Through Cloud and Intelligent Parking System. In *Handbook of Research on AI and ML for Intelligent Machines and Systems* (pp. 195–222). IGI Global.

Shneiderman. (2023). Bridging the gap between ethics and practice: Guidelines for Reliable, safe and trustworthy human-centred AI system. ACM Journals.

Sonia, R., Gupta, N., Manikandan, K., Hemalatha, R., Kumar, M. J., & Boopathi, S. (2024). Strengthening Security, Privacy, and Trust in Artificial Intelligence Drones for Smart Cities. In *Analyzing and Mitigating Security Risks in Cloud Computing* (pp. 214–242). IGI Global. doi:10.4018/979-8-3693-3249-8.ch011

Srinivas, B., Maguluri, L. P., Naidu, K. V., Reddy, L. C. S., Deivakani, M., & Boopathi, S. (2023). Architecture and Framework for Interfacing Cloud-Enabled Robots. In *Handbook of Research on Data Science and Cybersecurity Innovations in Industry 4.0 Technologies* (pp. 542–560). IGI Global. doi:10.4018/978-1-6684-8145-5.ch027

Stojkoska, B. L. (2017). A review of Internet of Things for smart home: Challenges and solutions. *Journal of Cleaner Production*, *140*, 1454-1464.

Subha, S., Inbamalar, T., Komala, C., Suresh, L. R., Boopathi, S., & Alaskar, K. (2023). A Remote Health Care Monitoring system using internet of medical things (IoMT). *IEEE Explore*, 1–6.

Suresh Babu, C. V. & Das, S. (2023). Impact of Blockchain Technology on the Stock Market. In K. Mehta, R. Sharma, & P. Yu (Eds.), *Revolutionizing Financial Services and Markets Through FinTech and Blockchain* (pp. 44-59). IGI Global. doi:10.4018/978-1-6684-8624-5.ch004

Suresh Babu, C. V., & Rohan, B. (2003). Evaluation and Quality Assurance for Rapid E-Learning and Development of Digital Learning Resources. In Implementing Rapid E-Learning Through Interactive Materials Development. IGI Global. doi:10.4018/978-1-6684-4940-0.ch008

Suresh Babu, C. V. (2022). *Artificial Intelligence and Expert Systems*. Anniyappa Publication.

Suresh Babu, C. V., & Padma, R. (2023). Technology Transformation Through Skilled Teachers in Teaching Accountancy. In R. González-Lezcano (Ed.), *Advancing STEM Education and Innovation in a Time of Distance Learning* (pp. 211–233). IGI Global. doi:10.4018/978-1-6684-5053-6.ch011

Suryadevara, C. K. (2023). Transforming Business Operations: Harnessing Artificial Intelligence and Machine Learning in the Enterprise. *International Journal of Creative Research Thoughts*, 2320–2882.

Tabassum, Z. (2023). Artificial intelligence and blockchain technology for secure smart grid and power distribution automation. In AI and Blockchain Applications in Industrial Robotics (pp. 226–252). IGI Global. doi:10.4018/979-8-3693-0659-8.ch009

Taguchi, G., & Shih-Chung, T. (1992). *Introduction to quality engineering: bringing quality engineering upstream*. ASME.

Tirlangi, S., Teotia, S., Padmapriya, G., Senthil Kumar, S., Dhotre, S., & Boopathi, S. (2024). Cloud Computing and Machine Learning in the Green Power Sector: Data Management and Analysis for Sustainable Energy. In Developments Towards Next Generation Intelligent Systems for Sustainable Development (pp. 148–179). IGI Global. doi:10.4018/979-8-3693-5643-2.ch006

Tong, Y., Liu, H., & Zhang, Z. (2024). Advancements in humanoid robots: A Comprehensive Review and future prospects. *IEEE/CAA Journal of Automatica Sinica, 11*(2), 301–328. doi:10.1109/JAS.2023.124140

Topol, E. J. (2019). High-performance medicine: The convergence of human and artificial intelligence. *Nature Medicine, 25*(1), 44–56. doi:10.1038/s41591-018-0300-7 PMID:30617339

Upadhyaya, A. N., Saqib, A., Devi, J. V., Rallapalli, S., Sudha, S., & Boopathi, S. (2024). Implementation of the Internet of Things (IoT) in Remote Healthcare. In Advances in Medical Technologies and Clinical Practice (pp. 104–124). IGI Global. doi:10.4018/979-8-3693-1934-5.ch006

Vaka, A. R., Soni, B., & Reddy, S. (2020). Breast cancer detection by leveraging Machine Learning. *Ict Express, 6*(4), 320–324. doi:10.1016/j.icte.2020.04.009

Veeranjaneyulu, R., Boopathi, S., Kumari, R. K., Vidyarthi, A., Isaac, J. S., & Jaiganesh, V. (2023). Air Quality Improvement and Optimisation Using Machine Learning Technique. *IEEE-Explore*, 1–6.

Vemuri, N. V. N. (2023). Enhancing Human-Robot Collaboration in Industry 4.0 with AI-driven HRI. *Power System Technology, 47*(4), 341–358. doi:10.52783/pst.196

Venkatasubramanian, V., Chitra, M., Sudha, R., Singh, V. P., Jefferson, K., & Boopathi, S. (2024). Examining the Impacts of Course Outcome Analysis in Indian Higher Education: Enhancing Educational Quality. In Challenges of Globalization and Inclusivity in Academic Research (pp. 124–145). IGI Global.

Venkataswamy, R., Janamala, V., & Cherukuri, R. C. (2023). Realization of humanoid doctor and real-time diagnostics of disease using internet of things, Edge Impulse Platform, and chatgpt. *Annals of Biomedical Engineering*, *52*(4), 738–740. doi:10.1007/s10439-023-03316-9 PMID:37453975

Venkateswaran, N., Vidhya, K., Ayyannan, M., Chavan, S. M., Sekar, K., & Boopathi, S. (2023). A Study on Smart Energy Management Framework Using Cloud Computing. In 5G, Artificial Intelligence, and Next Generation Internet of Things: Digital Innovation for Green and Sustainable Economies (pp. 189–212). IGI Global. doi:10.4018/978-1-6684-8634-4.ch009

Venkateswaran, N., Kumar, S. S., Diwakar, G., Gnanasangeetha, D., & Boopathi, S. (2023). Synthetic Biology for Waste Water to Energy Conversion: IoT and AI Approaches. *Applications of Synthetic Biology in Health. Energy & Environment*, 360–384.

Venkateswaran, N., Vidhya, R., Naik, D. A., Raj, T. M., Munjal, N., & Boopathi, S. (2023). Study on Sentence and Question Formation Using Deep Learning Techniques. In *Digital Natives as a Disruptive Force in Asian Businesses and Societies* (pp. 252–273). IGI Global. doi:10.4018/978-1-6684-6782-4.ch015

Vennila, T., Karuna, M., Srivastava, B. K., Venugopal, J., Surakasi, R., & Sampath, B. (2022). New Strategies in Treatment and Enzymatic Processes: Ethanol Production From Sugarcane Bagasse. In Human Agro-Energy Optimization for Business and Industry (pp. 219–240). IGI Global.

Verma, R., Christiana, M. B. V., Maheswari, M., Srinivasan, V., Patro, P., Dari, S. S., & Boopathi, S. (2024). Intelligent Physarum Solver for Profit Maximization in Oligopolistic Supply Chain Networks. In *AI and Machine Learning Impacts in Intelligent Supply Chain* (pp. 156–179). IGI Global. doi:10.4018/979-8-3693-1347-3.ch011

Vermesan, O., Bröring, A., Tragos, E., Serrano, M., Bacciu, D., Chessa, S., Gallicchio, C., Micheli, A., Dragone, M., Saffiotti, A., & ... (2022). Internet of robotic things–converging sensing/actuating, hyperconnectivity, artificial intelligence and IoT platforms. In *Cognitive Hyperconnected Digital Transformation* (pp. 97–155). River Publishers. doi:10.1201/9781003337584-4

Vignesh, S., Arulshri, K., SyedSajith, S., Kathiresan, S., Boopathi, S., & Dinesh Babu, P. (2018). Design and development of ornithopter and experimental analysis of flapping rate under various operating conditions. *Materials Today: Proceedings*, *5*(11), 25185–25194. doi:10.1016/j.matpr.2018.10.320

Vrontis, D., Christofi, M., Pereira, V., Tarba, S., Makrides, A., & Trichina, E. (2022). Artificial intelligence, robotics, advanced technologies and human resource management: A systematic review. *International Journal of Human Resource Management*, *33*(6), 1237–1266. Advance online publication. doi:10.1080/09585192.2020.1871398

Wada, K. (2003). Effects of robot assisted activity to elderly people who stay at a health service facility for the aged. *IEEE Int Conf Intell Robots Syst.*, *3*, 2847–2852.

Wasserman, G. (2003). *Reliability Verification, Testing, and Analysis in Engineering Design.* Marcel Dekker.

Wilhelm, J., Petzoldt, C., Beinke, T., & Freitag, M. (2021). Review of digital twin-based interaction in smart manufacturing: Enabling cyber-physical systems for human-machine interaction. *International Journal of Computer Integrated Manufacturing*, *34*(10), 1031–1048. doi:10.108 0/0951192X.2021.1963482

Woo, S., & O'Neal, D. L. (2021). Reliability Design of Mechanical Systems Such as Compressor Subjected to Repetitive Stresses. *Metals*, *11*(8), 1261. doi:10.3390/met11081261

Woo, S., O'Neal, D., & Pecht, M. (2023). Improving the lifetime of mechanical systems during transit established on quantum/transport life-stress prototype and sample size. *Mechanical Systems and Signal Processing*, *193*, 110222. doi:10.1016/j.ymssp.2023.110222

WRDA 2020 Updates. (2020). *The Final Report of the US House Committee on Transportation and Infrastructure on the Boeing 737 Max.* Available online: https://transportation.house.gov/committee-activity/boeing-737-max-investigation

Xie, D., Chen, L., Liu, L., Chen, L., & Wang, H. (2022). Actuators and Sensors for Application in Agricultural Robots: A Review. *Machines*, *2022*(10), 913. doi:10.3390/machines10100913

Xiong. (2024). Intuitive Human-Robot-Environment Interaction With EMG Signals: A Review. *IEEE/CAA Journal of Automatica Sinica.* . doi:10.1109/JAS.2024.124329

Yunjun Zheng. (2024). Adaptive Trajectory Tracking Control for Nonholonomic Wheeled Mobile Robots: A Barrier Function Sliding Mode Approach. *IEEE/CAA Journal of Automatica Sinica*, *11*(4), 1007-1021. . doi:10.1109/JAS.2023.124002

Yurtsever, E. (2020). A Survey of Autonomous Driving: Common Practices and Emerging Technologies. *IEEE Access, 8*, 58443–58469.

Zaidan, A. A., Zaidan, B. B., Qahtan, M. Y., Albahri, O. S., Albahri, A. S., Alaa, M., Jumaah, F. M., Talal, M., Tan, K. L., Shir, W. L., & Lim, C. K. (2018). A survey on communication components for IoT-based technologies in smart homes. *Telecommunication Systems*, *69*(1), 1–25. doi:10.1007/s11235-018-0430-8

Zaki, M. (2019). Digital transformation: Harnessing digital technologies for the next generation of services. *Journal of Services Marketing*, *33*(4), 429–435. doi:10.1108/JSM-01-2019-0034

Zhang, Y. M., Yang, Y.-P., Zhang, W., & Na, S.-J. (2020). Advanced welding manufacturing: A brief analysis and review of challenges and solutions. *Journal of Manufacturing Science and Engineering*, *142*(11), 110816. doi:10.1115/1.4047947

Related References

To continue our tradition of advancing information science and technology research, we have compiled a list of recommended IGI Global readings. These references will provide additional information and guidance to further enrich your knowledge and assist you with your own research and future publications.

Aasi, P., Rusu, L., & Vieru, D. (2017). The Role of Culture in IT Governance Five Focus Areas: A Literature Review. *International Journal of IT/Business Alignment and Governance, 8*(2), 42-61. https://doi.org/ doi:10.4018/IJITBAG.2017070103

Abdrabo, A. A. (2018). Egypt's Knowledge-Based Development: Opportunities, Challenges, and Future Possibilities. In A. Alraouf (Ed.), *Knowledge-Based Urban Development in the Middle East* (pp. 80–101). Hershey, PA: IGI Global. doi:10.4018/978-1-5225-3734-2.ch005

Abu Doush, I., & Alhami, I. (2018). Evaluating the Accessibility of Computer Laboratories, Libraries, and Websites in Jordanian Universities and Colleges. *International Journal of Information Systems and Social Change, 9*(2), 44–60. doi:10.4018/IJISSC.2018040104

Adegbore, A. M., Quadri, M. O., & Oyewo, O. R. (2018). A Theoretical Approach to the Adoption of Electronic Resource Management Systems (ERMS) in Nigerian University Libraries. In A. Tella & T. Kwanya (Eds.), *Handbook of Research on Managing Intellectual Property in Digital Libraries* (pp. 292–311). Hershey, PA: IGI Global. doi:10.4018/978-1-5225-3093-0.ch015

Afolabi, O. A. (2018). Myths and Challenges of Building an Effective Digital Library in Developing Nations: An African Perspective. In A. Tella & T. Kwanya (Eds.), *Handbook of Research on Managing Intellectual Property in Digital Libraries* (pp. 51–79). Hershey, PA: IGI Global. doi:10.4018/978-1-5225-3093-0.ch004

Agarwal, P., Kurian, R., & Gupta, R. K. (2022). Additive Manufacturing Feature Taxonomy and Placement of Parts in AM Enclosure. In S. Salunkhe, H. Hussein, & J. Davim (Eds.), *Applications of Artificial Intelligence in Additive Manufacturing* (pp. 138–176). IGI Global. https://doi.org/10.4018/978-1-7998-8516-0.ch007

Al-Alawi, A. I., Al-Hammam, A. H., Al-Alawi, S. S., & AlAlawi, E. I. (2021). The Adoption of E-Wallets: Current Trends and Future Outlook. In Y. Albastaki, A. Razzaque, & A. Sarea (Eds.), *Innovative Strategies for Implementing FinTech in Banking* (pp. 242–262). IGI Global. https://doi.org/10.4018/978-1-7998-3257-7.ch015

Alsharo, M. (2017). Attitudes Towards Cloud Computing Adoption in Emerging Economies. *International Journal of Cloud Applications and Computing*, 7(3), 44–58. doi:10.4018/IJCAC.2017070102

Amer, T. S., & Johnson, T. L. (2017). Information Technology Progress Indicators: Research Employing Psychological Frameworks. In A. Mesquita (Ed.), *Research Paradigms and Contemporary Perspectives on Human-Technology Interaction* (pp. 168–186). Hershey, PA: IGI Global. doi:10.4018/978-1-5225-1868-6.ch008

Andreeva, A., & Yolova, G. (2021). Liability in Labor Legislation: New Challenges Related to the Use of Artificial Intelligence. In B. Vassileva & M. Zwilling (Eds.), *Responsible AI and Ethical Issues for Businesses and Governments* (pp. 214–232). IGI Global. https://doi.org/10.4018/978-1-7998-4285-9.ch012

Anohah, E. (2017). Paradigm and Architecture of Computing Augmented Learning Management System for Computer Science Education. *International Journal of Online Pedagogy and Course Design*, 7(2), 60–70. doi:10.4018/IJOPCD.2017040105

Anohah, E., & Suhonen, J. (2017). Trends of Mobile Learning in Computing Education from 2006 to 2014: A Systematic Review of Research Publications. *International Journal of Mobile and Blended Learning*, 9(1), 16–33. doi:10.4018/IJMBL.2017010102

Arbaiza, C. S., Huerta, H. V., & Rodriguez, C. R. (2021). Contributions to the Technological Adoption Model for the Peruvian Agro-Export Sector. *International Journal of E-Adoption*, 13(1), 1–17. https://doi.org/10.4018/IJEA.2021010101

Bailey, E. K. (2017). Applying Learning Theories to Computer Technology Supported Instruction. In M. Grassetti & S. Brookby (Eds.), *Advancing Next-Generation Teacher Education through Digital Tools and Applications* (pp. 61–81). Hershey, PA: IGI Global. doi:10.4018/978-1-5225-0965-3.ch004

Baker, J. D. (2021). Introduction to Machine Learning as a New Methodological Framework for Performance Assessment. In M. Bocarnea, B. Winston, & D. Dean (Eds.), *Handbook of Research on Advancements in Organizational Data Collection and Measurements: Strategies for Addressing Attitudes, Beliefs, and Behaviors* (pp. 326–342). IGI Global. https://doi.org/10.4018/978-1-7998-7665-6.ch021

Banerjee, S., Sing, T. Y., Chowdhury, A. R., & Anwar, H. (2018). Let's Go Green: Towards a Taxonomy of Green Computing Enablers for Business Sustainability. In M. Khosrow-Pour (Ed.), *Green Computing Strategies for Competitive Advantage and Business Sustainability* (pp. 89–109). Hershey, PA: IGI Global. doi:10.4018/978-1-5225-5017-4.ch005

Basham, R. (2018). Information Science and Technology in Crisis Response and Management. In M. Khosrow-Pour, D.B.A. (Ed.), Encyclopedia of Information Science and Technology, Fourth Edition (pp. 1407-1418). Hershey, PA: IGI Global. doi:10.4018/978-1-5225-2255-3.ch121

Batyashe, T., & Iyamu, T. (2018). Architectural Framework for the Implementation of Information Technology Governance in Organisations. In M. Khosrow-Pour, D.B.A. (Ed.), Encyclopedia of Information Science and Technology, Fourth Edition (pp. 810-819). Hershey, PA: IGI Global. doi:10.4018/978-1-5225-2255-3.ch070

Bekleyen, N., & Çelik, S. (2017). Attitudes of Adult EFL Learners towards Preparing for a Language Test via CALL. In D. Tafazoli & M. Romero (Eds.), *Multiculturalism and Technology-Enhanced Language Learning* (pp. 214–229). Hershey, PA: IGI Global. doi:10.4018/978-1-5225-1882-2.ch013

Bergeron, F., Croteau, A., Uwizeyemungu, S., & Raymond, L. (2017). A Framework for Research on Information Technology Governance in SMEs. In S. De Haes & W. Van Grembergen (Eds.), *Strategic IT Governance and Alignment in Business Settings* (pp. 53–81). Hershey, PA: IGI Global. doi:10.4018/978-1-5225-0861-8.ch003

Bhardwaj, M., Shukla, N., & Sharma, A. (2021). Improvement and Reduction of Clustering Overhead in Mobile Ad Hoc Network With Optimum Stable Bunching Algorithm. In S. Kumar, M. Trivedi, P. Ranjan, & A. Punhani (Eds.), *Evolution of Software-Defined Networking Foundations for IoT and 5G Mobile Networks* (pp. 139–158). IGI Global. https://doi.org/10.4018/978-1-7998-4685-7.ch008

Bhatt, G. D., Wang, Z., & Rodger, J. A. (2017). Information Systems Capabilities and Their Effects on Competitive Advantages: A Study of Chinese Companies. *Information Resources Management Journal, 30*(3), 41–57. doi:10.4018/IRMJ.2017070103

Bhattacharya, A. (2021). Blockchain, Cybersecurity, and Industry 4.0. In A. Tyagi, G. Rekha, & N. Sreenath (Eds.), *Opportunities and Challenges for Blockchain Technology in Autonomous Vehicles* (pp. 210–244). IGI Global. https://doi.org/10.4018/978-1-7998-3295-9.ch013

Bhyan, P., Shrivastava, B., & Kumar, N. (2022). Requisite Sustainable Development Contemplating Buildings: Economic and Environmental Sustainability. In A. Hussain, K. Tiwari, & A. Gupta (Eds.), *Addressing Environmental Challenges Through Spatial Planning* (pp. 269–288). IGI Global. https://doi.org/10.4018/978-1-7998-8331-9.ch014

Boido, C., Davico, P., & Spallone, R. (2021). Digital Tools Aimed to Represent Urban Survey. In M. Khosrow-Pour D.B.A. (Ed.), *Encyclopedia of Information Science and Technology, Fifth Edition* (pp. 1181-1195). IGI Global. https://doi.org/10.4018/978-1-7998-3479-3.ch082

Borkar, P. S., Chanana, P. U., Atwal, S. K., Londe, T. G., & Dalal, Y. D. (2021). The Replacement of HMI (Human-Machine Interface) in Industry Using Single Interface Through IoT. In R. Raut & A. Mihovska (Eds.), *Examining the Impact of Deep Learning and IoT on Multi-Industry Applications* (pp. 195–208). IGI Global. https://doi.org/10.4018/978-1-7998-7511-6.ch011

Brahmane, A. V., & Krishna, C. B. (2021). Rider Chaotic Biography Optimization-driven Deep Stacked Auto-encoder for Big Data Classification Using Spark Architecture: Rider Chaotic Biography Optimization. *International Journal of Web Services Research*, 18(3), 42–62. https://doi.org/10.4018/ijwsr.2021070103

Burcoff, A., & Shamir, L. (2017). Computer Analysis of Pablo Picasso's Artistic Style. *International Journal of Art, Culture and Design Technologies*, 6(1), 1–18. doi:10.4018/IJACDT.2017010101

Byker, E. J. (2017). I Play I Learn: Introducing Technological Play Theory. In C. Martin & D. Polly (Eds.), *Handbook of Research on Teacher Education and Professional Development* (pp. 297–306). Hershey, PA: IGI Global. doi:10.4018/978-1-5225-1067-3.ch016

Calongne, C. M., Stricker, A. G., Truman, B., & Arenas, F. J. (2017). Cognitive Apprenticeship and Computer Science Education in Cyberspace: Reimagining the Past. In A. Stricker, C. Calongne, B. Truman, & F. Arenas (Eds.), *Integrating an Awareness of Selfhood and Society into Virtual Learning* (pp. 180–197). Hershey, PA: IGI Global. doi:10.4018/978-1-5225-2182-2.ch013

Carneiro, A. D. (2017). Defending Information Networks in Cyberspace: Some Notes on Security Needs. In M. Dawson, D. Kisku, P. Gupta, J. Sing, & W. Li (Eds.), Developing Next-Generation Countermeasures for Homeland Security Threat Prevention (pp. 354-375). Hershey, PA: IGI Global. https://doi.org/ doi:10.4018/978-1-5225-0703-1.ch016

Carvalho, W. F., & Zarate, L. (2021). Causal Feature Selection. In A. Azevedo & M. Santos (Eds.), *Integration Challenges for Analytics, Business Intelligence, and Data Mining* (pp. 145-160). IGI Global. https://doi.org/10.4018/978-1-7998-5781-5.ch007

Chase, J. P., & Yan, Z. (2017). Affect in Statistics Cognition. In *Assessing and Measuring Statistics Cognition in Higher Education Online Environments: Emerging Research and Opportunities* (pp. 144–187). Hershey, PA: IGI Global. doi:10.4018/978-1-5225-2420-5.ch005

Chatterjee, A., Roy, S., & Shrivastava, R. (2021). A Machine Learning Approach to Prevent Cancer. In G. Rani & P. Tiwari (Eds.), *Handbook of Research on Disease Prediction Through Data Analytics and Machine Learning* (pp. 112–141). IGI Global. https://doi.org/10.4018/978-1-7998-2742-9.ch007

Cifci, M. A. (2021). Optimizing WSNs for CPS Using Machine Learning Techniques. In A. Luhach & A. Elçi (Eds.), *Artificial Intelligence Paradigms for Smart Cyber-Physical Systems* (pp. 204–228). IGI Global. https://doi.org/10.4018/978-1-7998-5101-1.ch010

Cimermanova, I. (2017). Computer-Assisted Learning in Slovakia. In D. Tafazoli & M. Romero (Eds.), *Multiculturalism and Technology-Enhanced Language Learning* (pp. 252–270). Hershey, PA: IGI Global. doi:10.4018/978-1-5225-1882-2.ch015

Cipolla-Ficarra, F. V., & Cipolla-Ficarra, M. (2018). Computer Animation for Ingenious Revival. In F. Cipolla-Ficarra, M. Ficarra, M. Cipolla-Ficarra, A. Quiroga, J. Alma, & J. Carré (Eds.), *Technology-Enhanced Human Interaction in Modern Society* (pp. 159–181). Hershey, PA: IGI Global. doi:10.4018/978-1-5225-3437-2.ch008

Cockrell, S., Damron, T. S., Melton, A. M., & Smith, A. D. (2018). Offshoring IT. In M. Khosrow-Pour, D.B.A. (Ed.), Encyclopedia of Information Science and Technology, Fourth Edition (pp. 5476-5489). Hershey, PA: IGI Global. https://doi.org/ doi:10.4018/978-1-5225-2255-3.ch476

Coffey, J. W. (2018). Logic and Proof in Computer Science: Categories and Limits of Proof Techniques. In J. Horne (Ed.), *Philosophical Perceptions on Logic and Order* (pp. 218–240). Hershey, PA: IGI Global. doi:10.4018/978-1-5225-2443-4.ch007

Dale, M. (2017). Re-Thinking the Challenges of Enterprise Architecture Implementation. In M. Tavana (Ed.), *Enterprise Information Systems and the Digitalization of Business Functions* (pp. 205–221). Hershey, PA: IGI Global. doi:10.4018/978-1-5225-2382-6.ch009

Das, A., & Mohanty, M. N. (2021). An Useful Review on Optical Character Recognition for Smart Era Generation. In A. Tyagi (Ed.), *Multimedia and Sensory Input for Augmented, Mixed, and Virtual Reality* (pp. 1–41). IGI Global. https://doi.org/10.4018/978-1-7998-4703-8.ch001

Dash, A. K., & Mohapatra, P. (2021). A Survey on Prematurity Detection of Diabetic Retinopathy Based on Fundus Images Using Deep Learning Techniques. In S. Saxena & S. Paul (Eds.), *Deep Learning Applications in Medical Imaging* (pp. 140–155). IGI Global. https://doi.org/10.4018/978-1-7998-5071-7.ch006

De Maere, K., De Haes, S., & von Kutzschenbach, M. (2017). CIO Perspectives on Organizational Learning within the Context of IT Governance. *International Journal of IT/Business Alignment and Governance, 8*(1), 32-47. https://doi.org/doi:10.4018/IJITBAG.2017010103

Demir, K., Çaka, C., Yaman, N. D., İslamoğlu, H., & Kuzu, A. (2018). Examining the Current Definitions of Computational Thinking. In H. Ozcinar, G. Wong, & H. Ozturk (Eds.), *Teaching Computational Thinking in Primary Education* (pp. 36–64). Hershey, PA: IGI Global. doi:10.4018/978-1-5225-3200-2.ch003

Deng, X., Hung, Y., & Lin, C. D. (2017). Design and Analysis of Computer Experiments. In S. Saha, A. Mandal, A. Narasimhamurthy, S. V, & S. Sangam (Eds.), Handbook of Research on Applied Cybernetics and Systems Science (pp. 264-279). Hershey, PA: IGI Global. doi:10.4018/978-1-5225-2498-4.ch013

Denner, J., Martinez, J., & Thiry, H. (2017). Strategies for Engaging Hispanic/Latino Youth in the US in Computer Science. In Y. Rankin & J. Thomas (Eds.), *Moving Students of Color from Consumers to Producers of Technology* (pp. 24–48). Hershey, PA: IGI Global. doi:10.4018/978-1-5225-2005-4.ch002

Devi, A. (2017). Cyber Crime and Cyber Security: A Quick Glance. In R. Kumar, P. Pattnaik, & P. Pandey (Eds.), *Detecting and Mitigating Robotic Cyber Security Risks* (pp. 160–171). Hershey, PA: IGI Global. doi:10.4018/978-1-5225-2154-9.ch011

Dhaya, R., & Kanthavel, R. (2022). Futuristic Research Perspectives of IoT Platforms. In D. Jeya Mala (Ed.), *Integrating AI in IoT Analytics on the Cloud for Healthcare Applications* (pp. 258–275). IGI Global. doi:10.4018/978-1-7998-9132-1.ch015

Doyle, D. J., & Fahy, P. J. (2018). Interactivity in Distance Education and Computer-Aided Learning, With Medical Education Examples. In M. Khosrow-Pour, D.B.A. (Ed.), Encyclopedia of Information Science and Technology, Fourth Edition (pp. 5829-5840). Hershey, PA: IGI Global. https://doi.org/ doi:10.4018/978-1-5225-2255-3.ch507

Eklund, P. (2021). Reinforcement Learning in Social Media Marketing. In B. Christiansen & T. Škrinjarić (Eds.), *Handbook of Research on Applied AI for International Business and Marketing Applications* (pp. 30–48). IGI Global. https://doi.org/10.4018/978-1-7998-5077-9.ch003

El Ghandour, N., Benaissa, M., & Lebbah, Y. (2021). An Integer Linear Programming-Based Method for the Extraction of Ontology Alignment. *International Journal of Information Technology and Web Engineering, 16*(2), 25–44. https://doi.org/10.4018/IJITWE.2021040102

Elias, N. I., & Walker, T. W. (2017). Factors that Contribute to Continued Use of E-Training among Healthcare Professionals. In F. Topor (Ed.), *Handbook of Research on Individualism and Identity in the Globalized Digital Age* (pp. 403–429). Hershey, PA: IGI Global. doi:10.4018/978-1-5225-0522-8.ch018

Fisher, R. L. (2018). Computer-Assisted Indian Matrimonial Services. In M. Khosrow-Pour, D.B.A. (Ed.), Encyclopedia of Information Science and Technology, Fourth Edition (pp. 4136-4145). Hershey, PA: IGI Global. doi:10.4018/978-1-5225-2255-3.ch358

Galiautdinov, R. (2021). Nonlinear Filtering in Artificial Neural Network Applications in Business and Engineering. In Q. Do (Ed.), *Artificial Neural Network Applications in Business and Engineering* (pp. 1–23). IGI Global. https://doi.org/10.4018/978-1-7998-3238-6.ch001

Gardner-McCune, C., & Jimenez, Y. (2017). Historical App Developers: Integrating CS into K-12 through Cross-Disciplinary Projects. In Y. Rankin & J. Thomas (Eds.), *Moving Students of Color from Consumers to Producers of Technology* (pp. 85–112). Hershey, PA: IGI Global. doi:10.4018/978-1-5225-2005-4.ch005

Garg, P. K. (2021). The Internet of Things-Based Technologies. In S. Kumar, M. Trivedi, P. Ranjan, & A. Punhani (Eds.), *Evolution of Software-Defined Networking Foundations for IoT and 5G Mobile Networks* (pp. 37–65). IGI Global. https://doi.org/10.4018/978-1-7998-4685-7.ch003

Garg, T., & Bharti, M. (2021). Congestion Control Protocols for UWSNs. In N. Goyal, L. Sapra, & J. Sandhu (Eds.), *Energy-Efficient Underwater Wireless Communications and Networking* (pp. 85–100). IGI Global. https://doi.org/10.4018/978-1-7998-3640-7.ch006

Gauttier, S. (2021). A Primer on Q-Method and the Study of Technology. In M. Khosrow-Pour D.B.A. (Eds.), *Encyclopedia of Information Science and Technology, Fifth Edition* (pp. 1746-1756). IGI Global. https://doi.org/10.4018/978-1-7998-3479-3.ch120

Ghafele, R., & Gibert, B. (2018). Open Growth: The Economic Impact of Open Source Software in the USA. In M. Khosrow-Pour (Ed.), *Optimizing Contemporary Application and Processes in Open Source Software* (pp. 164–197). Hershey, PA: IGI Global. doi:10.4018/978-1-5225-5314-4.ch007

Ghobakhloo, M., & Azar, A. (2018). Information Technology Resources, the Organizational Capability of Lean-Agile Manufacturing, and Business Performance. *Information Resources Management Journal*, 31(2), 47–74. doi:10.4018/IRMJ.2018040103

Gikandi, J. W. (2017). Computer-Supported Collaborative Learning and Assessment: A Strategy for Developing Online Learning Communities in Continuing Education. In J. Keengwe & G. Onchwari (Eds.), *Handbook of Research on Learner-Centered Pedagogy in Teacher Education and Professional Development* (pp. 309–333). Hershey, PA: IGI Global. doi:10.4018/978-1-5225-0892-2.ch017

Gokhale, A. A., & Machina, K. F. (2017). Development of a Scale to Measure Attitudes toward Information Technology. In L. Tomei (Ed.), *Exploring the New Era of Technology-Infused Education* (pp. 49–64). Hershey, PA: IGI Global. doi:10.4018/978-1-5225-1709-2.ch004

Goswami, J. K., Jalal, S., Negi, C. S., & Jalal, A. S. (2022). A Texture Features-Based Robust Facial Expression Recognition. *International Journal of Computer Vision and Image Processing*, 12(1), 1–15. https://doi.org/10.4018/IJCVIP.2022010103

Hafeez-Baig, A., Gururajan, R., & Wickramasinghe, N. (2017). Readiness as a Novel Construct of Readiness Acceptance Model (RAM) for the Wireless Handheld Technology. In N. Wickramasinghe (Ed.), *Handbook of Research on Healthcare Administration and Management* (pp. 578–595). Hershey, PA: IGI Global. doi:10.4018/978-1-5225-0920-2.ch035

Hanafizadeh, P., Ghandchi, S., & Asgarimehr, M. (2017). Impact of Information Technology on Lifestyle: A Literature Review and Classification. *International Journal of Virtual Communities and Social Networking, 9*(2), 1–23. doi:10.4018/IJVCSN.2017040101

Haseski, H. İ., Ilic, U., & Tuğtekin, U. (2018). Computational Thinking in Educational Digital Games: An Assessment Tool Proposal. In H. Ozcinar, G. Wong, & H. Ozturk (Eds.), *Teaching Computational Thinking in Primary Education* (pp. 256–287). Hershey, PA: IGI Global. doi:10.4018/978-1-5225-3200-2.ch013

Hee, W. J., Jalleh, G., Lai, H., & Lin, C. (2017). E-Commerce and IT Projects: Evaluation and Management Issues in Australian and Taiwanese Hospitals. *International Journal of Public Health Management and Ethics, 2*(1), 69–90. doi:10.4018/IJPHME.2017010104

Hernandez, A. A. (2017). Green Information Technology Usage: Awareness and Practices of Philippine IT Professionals. *International Journal of Enterprise Information Systems, 13*(4), 90–103. doi:10.4018/IJEIS.2017100106

Hernandez, M. A., Marin, E. C., Garcia-Rodriguez, J., Azorin-Lopez, J., & Cazorla, M. (2017). Automatic Learning Improves Human-Robot Interaction in Productive Environments: A Review. *International Journal of Computer Vision and Image Processing, 7*(3), 65–75. doi:10.4018/IJCVIP.2017070106

Hirota, A. (2021). Design of Narrative Creation in Innovation: "Signature Story" and Two Types of Pivots. In T. Ogata & J. Ono (Eds.), *Bridging the Gap Between AI, Cognitive Science, and Narratology With Narrative Generation* (pp. 363–376). IGI Global. https://doi.org/10.4018/978-1-7998-4864-6.ch012

Hond, D., Asgari, H., Jeffery, D., & Newman, M. (2021). An Integrated Process for Verifying Deep Learning Classifiers Using Dataset Dissimilarity Measures. *International Journal of Artificial Intelligence and Machine Learning, 11*(2), 1–21. https://doi.org/10.4018/IJAIML.289536

Horne-Popp, L. M., Tessone, E. B., & Welker, J. (2018). If You Build It, They Will Come: Creating a Library Statistics Dashboard for Decision-Making. In L. Costello & M. Powers (Eds.), *Developing In-House Digital Tools in Library Spaces* (pp. 177–203). Hershey, PA: IGI Global. doi:10.4018/978-1-5225-2676-6.ch009

Hu, H., Hu, P. J., & Al-Gahtani, S. S. (2017). User Acceptance of Computer Technology at Work in Arabian Culture: A Model Comparison Approach. In M. Khosrow-Pour (Ed.), *Handbook of Research on Technology Adoption, Social Policy, and Global Integration* (pp. 205–228). Hershey, PA: IGI Global. doi:10.4018/978-1-5225-2668-1.ch011

Huang, C., Sun, Y., & Fuh, C. (2022). Vehicle License Plate Recognition With Deep Learning. In C. Chen, W. Yang, & L. Chen (Eds.), *Technologies to Advance Automation in Forensic Science and Criminal Investigation* (pp. 161-219). IGI Global. https://doi.org/10.4018/978-1-7998-8386-9.ch009

Ifinedo, P. (2017). Using an Extended Theory of Planned Behavior to Study Nurses' Adoption of Healthcare Information Systems in Nova Scotia. *International Journal of Technology Diffusion*, 8(1), 1–17. doi:10.4018/IJTD.2017010101

Ilie, V., & Sneha, S. (2018). A Three Country Study for Understanding Physicians' Engagement With Electronic Information Resources Pre and Post System Implementation. *Journal of Global Information Management*, 26(2), 48–73. doi:10.4018/JGIM.2018040103

Ilo, P. I., Nkiko, C., Ugwu, C. I., Ekere, J. N., Izuagbe, R., & Fagbohun, M. O. (2021). Prospects and Challenges of Web 3.0 Technologies Application in the Provision of Library Services. In M. Khosrow-Pour D.B.A. (Ed.), *Encyclopedia of Information Science and Technology, Fifth Edition* (pp. 1767-1781). IGI Global. https://doi.org/10.4018/978-1-7998-3479-3.ch122

Inoue-Smith, Y. (2017). Perceived Ease in Using Technology Predicts Teacher Candidates' Preferences for Online Resources. *International Journal of Online Pedagogy and Course Design*, 7(3), 17–28. doi:10.4018/IJOPCD.2017070102

Islam, A. Y. (2017). Technology Satisfaction in an Academic Context: Moderating Effect of Gender. In A. Mesquita (Ed.), *Research Paradigms and Contemporary Perspectives on Human-Technology Interaction* (pp. 187–211). Hershey, PA: IGI Global. doi:10.4018/978-1-5225-1868-6.ch009

Jagdale, S. C., Hable, A. A., & Chabukswar, A. R. (2021). Protocol Development in Clinical Trials for Healthcare Management. In M. Khosrow-Pour D.B.A. (Ed.), *Encyclopedia of Information Science and Technology, Fifth Edition* (pp. 1797-1814). IGI Global. https://doi.org/10.4018/978-1-7998-3479-3.ch124

Jamil, G. L., & Jamil, C. C. (2017). Information and Knowledge Management Perspective Contributions for Fashion Studies: Observing Logistics and Supply Chain Management Processes. In G. Jamil, A. Soares, & C. Pessoa (Eds.), *Handbook of Research on Information Management for Effective Logistics and Supply Chains* (pp. 199–221). Hershey, PA: IGI Global. doi:10.4018/978-1-5225-0973-8.ch011

Jamil, M. I., & Almunawar, M. N. (2021). Importance of Digital Literacy and Hindrance Brought About by Digital Divide. In M. Khosrow-Pour D.B.A. (Ed.), *Encyclopedia of Information Science and Technology, Fifth Edition* (pp. 1683-1698). IGI Global. https://doi.org/10.4018/978-1-7998-3479-3.ch116

Janakova, M. (2018). Big Data and Simulations for the Solution of Controversies in Small Businesses. In M. Khosrow-Pour, D.B.A. (Ed.), Encyclopedia of Information Science and Technology, Fourth Edition (pp. 6907-6915). Hershey, PA: IGI Global. doi:10.4018/978-1-5225-2255-3.ch598

Jhawar, A., & Garg, S. K. (2018). Logistics Improvement by Investment in Information Technology Using System Dynamics. In A. Azar & S. Vaidyanathan (Eds.), *Advances in System Dynamics and Control* (pp. 528–567). Hershey, PA: IGI Global. doi:10.4018/978-1-5225-4077-9.ch017

Kalelioğlu, F., Gülbahar, Y., & Doğan, D. (2018). Teaching How to Think Like a Programmer: Emerging Insights. In H. Ozcinar, G. Wong, & H. Ozturk (Eds.), *Teaching Computational Thinking in Primary Education* (pp. 18–35). Hershey, PA: IGI Global. doi:10.4018/978-1-5225-3200-2.ch002

Kamberi, S. (2017). A Girls-Only Online Virtual World Environment and its Implications for Game-Based Learning. In A. Stricker, C. Calongne, B. Truman, & F. Arenas (Eds.), *Integrating an Awareness of Selfhood and Society into Virtual Learning* (pp. 74–95). Hershey, PA: IGI Global. doi:10.4018/978-1-5225-2182-2.ch006

Kamel, S., & Rizk, N. (2017). ICT Strategy Development: From Design to Implementation – Case of Egypt. In C. Howard & K. Hargiss (Eds.), *Strategic Information Systems and Technologies in Modern Organizations* (pp. 239–257). Hershey, PA: IGI Global. doi:10.4018/978-1-5225-1680-4.ch010

Kamel, S. H. (2018). The Potential Role of the Software Industry in Supporting Economic Development. In M. Khosrow-Pour, D.B.A. (Ed.), Encyclopedia of Information Science and Technology, Fourth Edition (pp. 7259-7269). Hershey, PA: IGI Global. doi:10.4018/978-1-5225-2255-3.ch631

Kang, H., Kang, Y., & Kim, J. (2022). Improved Fall Detection Model on GRU Using PoseNet. *International Journal of Software Innovation, 10*(2), 1–11. https://doi.org/10.4018/IJSI.289600

Kankam, P. K. (2021). Employing Case Study and Survey Designs in Information Research. *Journal of Information Technology Research, 14*(1), 167–177. https://doi.org/10.4018/JITR.2021010110

Karas, V., & Schuller, B. W. (2021). Deep Learning for Sentiment Analysis: An Overview and Perspectives. In F. Pinarbasi & M. Taskiran (Eds.), *Natural Language Processing for Global and Local Business* (pp. 97–132). IGI Global. https://doi.org/10.4018/978-1-7998-4240-8.ch005

Kaufman, L. M. (2022). Reimagining the Magic of the Workshop Model. In T. Driscoll III, (Ed.), *Designing Effective Distance and Blended Learning Environments in K-12* (pp. 89–109). IGI Global. https://doi.org/10.4018/978-1-7998-6829-3.ch007

Kawata, S. (2018). Computer-Assisted Parallel Program Generation. In M. Khosrow-Pour, D.B.A. (Ed.), Encyclopedia of Information Science and Technology, Fourth Edition (pp. 4583-4593). Hershey, PA: IGI Global. doi:10.4018/978-1-5225-2255-3.ch398

Kharb, L., & Singh, P. (2021). Role of Machine Learning in Modern Education and Teaching. In S. Verma & P. Tomar (Ed.), *Impact of AI Technologies on Teaching, Learning, and Research in Higher Education* (pp. 99-123). IGI Global. https://doi.org/10.4018/978-1-7998-4763-2.ch006

Khari, M., Shrivastava, G., Gupta, S., & Gupta, R. (2017). Role of Cyber Security in Today's Scenario. In R. Kumar, P. Pattnaik, & P. Pandey (Eds.), *Detecting and Mitigating Robotic Cyber Security Risks* (pp. 177–191). Hershey, PA: IGI Global. doi:10.4018/978-1-5225-2154-9.ch013

Khekare, G., & Sheikh, S. (2021). Autonomous Navigation Using Deep Reinforcement Learning in ROS. *International Journal of Artificial Intelligence and Machine Learning*, *11*(2), 63–70. https://doi.org/10.4018/IJAIML.20210701.oa4

Khouja, M., Rodriguez, I. B., Ben Halima, Y., & Moalla, S. (2018). IT Governance in Higher Education Institutions: A Systematic Literature Review. *International Journal of Human Capital and Information Technology Professionals*, *9*(2), 52–67. doi:10.4018/IJHCITP.2018040104

Kiourt, C., Pavlidis, G., Koutsoudis, A., & Kalles, D. (2017). Realistic Simulation of Cultural Heritage. *International Journal of Computational Methods in Heritage Science*, *1*(1), 10–40. doi:10.4018/IJCMHS.2017010102

Köse, U. (2017). An Augmented-Reality-Based Intelligent Mobile Application for Open Computer Education. In G. Kurubacak & H. Altinpulluk (Eds.), *Mobile Technologies and Augmented Reality in Open Education* (pp. 154–174). Hershey, PA: IGI Global. doi:10.4018/978-1-5225-2110-5.ch008

Lahmiri, S. (2018). Information Technology Outsourcing Risk Factors and Provider Selection. In M. Gupta, R. Sharman, J. Walp, & P. Mulgund (Eds.), *Information Technology Risk Management and Compliance in Modern Organizations* (pp. 214–228). Hershey, PA: IGI Global. doi:10.4018/978-1-5225-2604-9.ch008

Lakkad, A. K., Bhadaniya, R. D., Shah, V. N., & Lavanya, K. (2021). Complex Events Processing on Live News Events Using Apache Kafka and Clustering Techniques. *International Journal of Intelligent Information Technologies*, *17*(1), 39–52. https://doi.org/10.4018/IJIIT.2021010103

Landriscina, F. (2017). Computer-Supported Imagination: The Interplay Between Computer and Mental Simulation in Understanding Scientific Concepts. In I. Levin & D. Tsybulsky (Eds.), *Digital Tools and Solutions for Inquiry-Based STEM Learning* (pp. 33–60). Hershey, PA: IGI Global. doi:10.4018/978-1-5225-2525-7.ch002

Lara López, G. (2021). Virtual Reality in Object Location. In A. Negrón & M. Muñoz (Eds.), *Latin American Women and Research Contributions to the IT Field* (pp. 307–324). IGI Global. https://doi.org/10.4018/978-1-7998-7552-9.ch014

Lee, W. W. (2018). Ethical Computing Continues From Problem to Solution. In M. Khosrow-Pour, D.B.A. (Ed.), Encyclopedia of Information Science and Technology, Fourth Edition (pp. 4884-4897). Hershey, PA: IGI Global. doi:10.4018/978-1-5225-2255-3.ch423

Lin, S., Chen, S., & Chuang, S. (2017). Perceived Innovation and Quick Response Codes in an Online-to-Offline E-Commerce Service Model. *International Journal of E-Adoption*, *9*(2), 1–16. doi:10.4018/IJEA.2017070101

Liu, M., Wang, Y., Xu, W., & Liu, L. (2017). Automated Scoring of Chinese Engineering Students' English Essays. *International Journal of Distance Education Technologies*, *15*(1), 52–68. doi:10.4018/IJDET.2017010104

Ma, X., Li, X., Zhong, B., Huang, Y., Gu, Y., Wu, M., Liu, Y., & Zhang, M. (2021). A Detector and Evaluation Framework of Abnormal Bidding Behavior Based on Supplier Portrait. *International Journal of Information Technology and Web Engineering*, *16*(2), 58–74. https://doi.org/10.4018/IJITWE.2021040104

Mabe, L. K., & Oladele, O. I. (2017). Application of Information Communication Technologies for Agricultural Development through Extension Services: A Review. In T. Tossy (Ed.), *Information Technology Integration for Socio-Economic Development* (pp. 52–101). Hershey, PA: IGI Global. doi:10.4018/978-1-5225-0539-6.ch003

Mahboub, S. A., Sayed Ali Ahmed, E., & Saeed, R. A. (2021). Smart IDS and IPS for Cyber-Physical Systems. In A. Luhach & A. Elçi (Eds.), *Artificial Intelligence Paradigms for Smart Cyber-Physical Systems* (pp. 109–136). IGI Global. https://doi.org/10.4018/978-1-7998-5101-1.ch006

Manogaran, G., Thota, C., & Lopez, D. (2018). Human-Computer Interaction With Big Data Analytics. In D. Lopez & M. Durai (Eds.), *HCI Challenges and Privacy Preservation in Big Data Security* (pp. 1–22). Hershey, PA: IGI Global. doi:10.4018/978-1-5225-2863-0.ch001

Margolis, J., Goode, J., & Flapan, J. (2017). A Critical Crossroads for Computer Science for All: "Identifying Talent" or "Building Talent," and What Difference Does It Make? In Y. Rankin & J. Thomas (Eds.), *Moving Students of Color from Consumers to Producers of Technology* (pp. 1–23). Hershey, PA: IGI Global. doi:10.4018/978-1-5225-2005-4.ch001

Mazzù, M. F., Benetton, A., Baccelloni, A., & Lavini, L. (2022). A Milk Blockchain-Enabled Supply Chain: Evidence From Leading Italian Farms. In P. De Giovanni (Ed.), *Blockchain Technology Applications in Businesses and Organizations* (pp. 73–98). IGI Global. https://doi.org/10.4018/978-1-7998-8014-1.ch004

Mbale, J. (2018). Computer Centres Resource Cloud Elasticity-Scalability (CRECES): Copperbelt University Case Study. In S. Aljawarneh & M. Malhotra (Eds.), *Critical Research on Scalability and Security Issues in Virtual Cloud Environments* (pp. 48–70). Hershey, PA: IGI Global. doi:10.4018/978-1-5225-3029-9.ch003

McKee, J. (2018). The Right Information: The Key to Effective Business Planning. In *Business Architectures for Risk Assessment and Strategic Planning: Emerging Research and Opportunities* (pp. 38–52). Hershey, PA: IGI Global. doi:10.4018/978-1-5225-3392-4.ch003

Meddah, I. H., Remil, N. E., & Meddah, H. N. (2021). Novel Approach for Mining Patterns. *International Journal of Applied Evolutionary Computation, 12*(1), 27–42. https://doi.org/10.4018/IJAEC.2021010103

Mensah, I. K., & Mi, J. (2018). Determinants of Intention to Use Local E-Government Services in Ghana: The Perspective of Local Government Workers. *International Journal of Technology Diffusion, 9*(2), 41–60. doi:10.4018/IJTD.2018040103

Mohamed, J. H. (2018). Scientograph-Based Visualization of Computer Forensics Research Literature. In J. Jeyasekar & P. Saravanan (Eds.), *Innovations in Measuring and Evaluating Scientific Information* (pp. 148–162). Hershey, PA: IGI Global. doi:10.4018/978-1-5225-3457-0.ch010

Montañés-Del Río, M. Á., Cornejo, V. R., Rodríguez, M. R., & Ortiz, J. S. (2021). Gamification of University Subjects: A Case Study for Operations Management. *Journal of Information Technology Research, 14*(2), 1–29. https://doi.org/10.4018/JITR.2021040101

Moore, R. L., & Johnson, N. (2017). Earning a Seat at the Table: How IT Departments Can Partner in Organizational Change and Innovation. *International Journal of Knowledge-Based Organizations, 7*(2), 1–12. doi:10.4018/IJKBO.2017040101

Mukul, M. K., & Bhattaharyya, S. (2017). Brain-Machine Interface: Human-Computer Interaction. In E. Noughabi, B. Raahemi, A. Albadvi, & B. Far (Eds.), *Handbook of Research on Data Science for Effective Healthcare Practice and Administration* (pp. 417–443). Hershey, PA: IGI Global. doi:10.4018/978-1-5225-2515-8.ch018

Na, L. (2017). Library and Information Science Education and Graduate Programs in Academic Libraries. In L. Ruan, Q. Zhu, & Y. Ye (Eds.), *Academic Library Development and Administration in China* (pp. 218–229). Hershey, PA: IGI Global. doi:10.4018/978-1-5225-0550-1.ch013

Nagpal, G., Bishnoi, G. K., Dhami, H. S., & Vijayvargia, A. (2021). Use of Data Analytics to Increase the Efficiency of Last Mile Logistics for Ecommerce Deliveries. In B. Patil & M. Vohra (Eds.), *Handbook of Research on Engineering, Business, and Healthcare Applications of Data Science and Analytics* (pp. 167–180). IGI Global. https://doi.org/10.4018/978-1-7998-3053-5.ch009

Nair, S. M., Ramesh, V., & Tyagi, A. K. (2021). Issues and Challenges (Privacy, Security, and Trust) in Blockchain-Based Applications. In A. Tyagi, G. Rekha, & N. Sreenath (Eds.), *Opportunities and Challenges for Blockchain Technology in Autonomous Vehicles* (pp. 196–209). IGI Global. https://doi.org/10.4018/978-1-7998-3295-9.ch012

Naomi, J. F. M., K., & V., S. (2021). Machine and Deep Learning Techniques in IoT and Cloud. In S. Velayutham (Ed.), *Challenges and Opportunities for the Convergence of IoT, Big Data, and Cloud Computing* (pp. 225-247). IGI Global. https://doi.org/10.4018/978-1-7998-3111-2.ch013

Nath, R., & Murthy, V. N. (2018). What Accounts for the Differences in Internet Diffusion Rates Around the World? In M. Khosrow-Pour, D.B.A. (Ed.), Encyclopedia of Information Science and Technology, Fourth Edition (pp. 8095-8104). Hershey, PA: IGI Global. https://doi.org/ doi:10.4018/978-1-5225-2255-3.ch705

Nedelko, Z., & Potocan, V. (2018). The Role of Emerging Information Technologies for Supporting Supply Chain Management. In M. Khosrow-Pour, D.B.A. (Ed.), Encyclopedia of Information Science and Technology, Fourth Edition (pp. 5559-5569). Hershey, PA: IGI Global. doi:10.4018/978-1-5225-2255-3.ch483

Negrini, L., Giang, C., & Bonnet, E. (2022). Designing Tools and Activities for Educational Robotics in Online Learning. In N. Eteokleous & E. Nisiforou (Eds.), *Designing, Constructing, and Programming Robots for Learning* (pp. 202–222). IGI Global. https://doi.org/10.4018/978-1-7998-7443-0.ch010

Ngafeeson, M. N. (2018). User Resistance to Health Information Technology. In M. Khosrow-Pour, D.B.A. (Ed.), Encyclopedia of Information Science and Technology, Fourth Edition (pp. 3816-3825). Hershey, PA: IGI Global. doi:10.4018/978-1-5225-2255-3.ch331

Nguyen, T. T., Giang, N. L., Tran, D. T., Nguyen, T. T., Nguyen, H. Q., Pham, A. V., & Vu, T. D. (2021). A Novel Filter-Wrapper Algorithm on Intuitionistic Fuzzy Set for Attribute Reduction From Decision Tables. *International Journal of Data Warehousing and Mining*, *17*(4), 67–100. https://doi.org/10.4018/IJDWM.2021100104

Nigam, A., & Dewani, P. P. (2022). Consumer Engagement Through Conditional Promotions: An Exploratory Study. *Journal of Global Information Management*, *30*(5), 1–19. https://doi.org/10.4018/JGIM.290364

Odagiri, K. (2017). Introduction of Individual Technology to Constitute the Current Internet. In *Strategic Policy-Based Network Management in Contemporary Organizations* (pp. 20–96). Hershey, PA: IGI Global. doi:10.4018/978-1-68318-003-6.ch003

Odia, J. O., & Akpata, O. T. (2021). Role of Data Science and Data Analytics in Forensic Accounting and Fraud Detection. In B. Patil & M. Vohra (Eds.), *Handbook of Research on Engineering, Business, and Healthcare Applications of Data Science and Analytics* (pp. 203–227). IGI Global. https://doi.org/10.4018/978-1-7998-3053-5.ch011

Okike, E. U. (2018). Computer Science and Prison Education. In I. Biao (Ed.), *Strategic Learning Ideologies in Prison Education Programs* (pp. 246–264). Hershey, PA: IGI Global. doi:10.4018/978-1-5225-2909-5.ch012

Olelewe, C. J., & Nwafor, I. P. (2017). Level of Computer Appreciation Skills Acquired for Sustainable Development by Secondary School Students in Nsukka LGA of Enugu State, Nigeria. In C. Ayo & V. Mbarika (Eds.), *Sustainable ICT Adoption and Integration for Socio-Economic Development* (pp. 214–233). Hershey, PA: IGI Global. doi:10.4018/978-1-5225-2565-3.ch010

Oliveira, M., Maçada, A. C., Curado, C., & Nodari, F. (2017). Infrastructure Profiles and Knowledge Sharing. *International Journal of Technology and Human Interaction*, *13*(3), 1–12. doi:10.4018/IJTHI.2017070101

Otarkhani, A., Shokouhyar, S., & Pour, S. S. (2017). Analyzing the Impact of Governance of Enterprise IT on Hospital Performance: Tehran's (Iran) Hospitals – A Case Study. *International Journal of Healthcare Information Systems and Informatics*, *12*(3), 1–20. doi:10.4018/IJHISI.2017070101

Otunla, A. O., & Amuda, C. O. (2018). Nigerian Undergraduate Students' Computer Competencies and Use of Information Technology Tools and Resources for Study Skills and Habits' Enhancement. In M. Khosrow-Pour, D.B.A. (Ed.), Encyclopedia of Information Science and Technology, Fourth Edition (pp. 2303-2313). Hershey, PA: IGI Global. https://doi.org/ doi:10.4018/978-1-5225-2255-3.ch200

Özçınar, H. (2018). A Brief Discussion on Incentives and Barriers to Computational Thinking Education. In H. Ozcinar, G. Wong, & H. Ozturk (Eds.), *Teaching Computational Thinking in Primary Education* (pp. 1–17). Hershey, PA: IGI Global. doi:10.4018/978-1-5225-3200-2.ch001

Pandey, J. M., Garg, S., Mishra, P., & Mishra, B. P. (2017). Computer Based Psychological Interventions: Subject to the Efficacy of Psychological Services. *International Journal of Computers in Clinical Practice*, *2*(1), 25–33. doi:10.4018/IJCCP.2017010102

Pandkar, S. D., & Paatil, S. D. (2021). Big Data and Knowledge Resource Centre. In S. Dhamdhere (Ed.), *Big Data Applications for Improving Library Services* (pp. 90–106). IGI Global. https://doi.org/10.4018/978-1-7998-3049-8.ch007

Patro, C. (2017). Impulsion of Information Technology on Human Resource Practices. In P. Ordóñez de Pablos (Ed.), *Managerial Strategies and Solutions for Business Success in Asia* (pp. 231–254). Hershey, PA: IGI Global. doi:10.4018/978-1-5225-1886-0.ch013

Patro, C. S., & Raghunath, K. M. (2017). Information Technology Paraphernalia for Supply Chain Management Decisions. In M. Tavana (Ed.), *Enterprise Information Systems and the Digitalization of Business Functions* (pp. 294–320). Hershey, PA: IGI Global. doi:10.4018/978-1-5225-2382-6.ch014

Paul, P. K. (2018). The Context of IST for Solid Information Retrieval and Infrastructure Building: Study of Developing Country. *International Journal of Information Retrieval Research*, *8*(1), 86–100. doi:10.4018/IJIRR.2018010106

Paul, P. K., & Chatterjee, D. (2018). iSchools Promoting "Information Science and Technology" (IST) Domain Towards Community, Business, and Society With Contemporary Worldwide Trend and Emerging Potentialities in India. In M. Khosrow-Pour, D.B.A. (Ed.), Encyclopedia of Information Science and Technology, Fourth Edition (pp. 4723-4735). Hershey, PA: IGI Global. https://doi.org/ doi:10.4018/978-1-5225-2255-3.ch410

Pessoa, C. R., & Marques, M. E. (2017). Information Technology and Communication Management in Supply Chain Management. In G. Jamil, A. Soares, & C. Pessoa (Eds.), *Handbook of Research on Information Management for Effective Logistics and Supply Chains* (pp. 23–33). Hershey, PA: IGI Global. doi:10.4018/978-1-5225-0973-8.ch002

Pineda, R. G. (2018). Remediating Interaction: Towards a Philosophy of Human-Computer Relationship. In M. Khosrow-Pour (Ed.), *Enhancing Art, Culture, and Design With Technological Integration* (pp. 75–98). Hershey, PA: IGI Global. doi:10.4018/978-1-5225-5023-5.ch004

Prabha, V. D., & R., R. (2021). Clinical Decision Support Systems: Decision-Making System for Clinical Data. In G. Rani & P. Tiwari (Eds.), *Handbook of Research on Disease Prediction Through Data Analytics and Machine Learning* (pp. 268-280). IGI Global. https://doi.org/10.4018/978-1-7998-2742-9.ch014

Pushpa, R., & Siddappa, M. (2021). An Optimal Way of VM Placement Strategy in Cloud Computing Platform Using ABCS Algorithm. *International Journal of Ambient Computing and Intelligence*, *12*(3), 16–38. https://doi.org/10.4018/IJACI.2021070102

Qian, Y. (2017). Computer Simulation in Higher Education: Affordances, Opportunities, and Outcomes. In P. Vu, S. Fredrickson, & C. Moore (Eds.), *Handbook of Research on Innovative Pedagogies and Technologies for Online Learning in Higher Education* (pp. 236–262). Hershey, PA: IGI Global. doi:10.4018/978-1-5225-1851-8.ch011

Rahman, N. (2017). Lessons from a Successful Data Warehousing Project Management. *International Journal of Information Technology Project Management*, *8*(4), 30–45. doi:10.4018/IJITPM.2017100103

Rahman, N. (2018). Environmental Sustainability in the Computer Industry for Competitive Advantage. In M. Khosrow-Pour (Ed.), *Green Computing Strategies for Competitive Advantage and Business Sustainability* (pp. 110–130). Hershey, PA: IGI Global. doi:10.4018/978-1-5225-5017-4.ch006

Rajh, A., & Pavetic, T. (2017). Computer Generated Description as the Required Digital Competence in Archival Profession. *International Journal of Digital Literacy and Digital Competence*, 8(1), 36–49. doi:10.4018/IJDLDC.2017010103

Raman, A., & Goyal, D. P. (2017). Extending IMPLEMENT Framework for Enterprise Information Systems Implementation to Information System Innovation. In M. Tavana (Ed.), *Enterprise Information Systems and the Digitalization of Business Functions* (pp. 137–177). Hershey, PA: IGI Global. doi:10.4018/978-1-5225-2382-6.ch007

Rao, A. P., & Reddy, K. S. (2021). Automated Soil Residue Levels Detecting Device With IoT Interface. In V. Sathiyamoorthi & A. Elci (Eds.), *Challenges and Applications of Data Analytics in Social Perspectives* (Vol. S, pp. 123–135). IGI Global. https://doi.org/10.4018/978-1-7998-2566-1.ch007

Rao, Y. S., Rauta, A. K., Saini, H., & Panda, T. C. (2017). Mathematical Model for Cyber Attack in Computer Network. *International Journal of Business Data Communications and Networking*, 13(1), 58–65. doi:10.4018/IJBDCN.2017010105

Rapaport, W. J. (2018). Syntactic Semantics and the Proper Treatment of Computationalism. In M. Danesi (Ed.), *Empirical Research on Semiotics and Visual Rhetoric* (pp. 128–176). Hershey, PA: IGI Global. doi:10.4018/978-1-5225-5622-0.ch007

Raut, R., Priyadarshinee, P., & Jha, M. (2017). Understanding the Mediation Effect of Cloud Computing Adoption in Indian Organization: Integrating TAM-TOE- Risk Model. *International Journal of Service Science, Management, Engineering, and Technology*, 8(3), 40–59. doi:10.4018/IJSSMET.2017070103

Rezaie, S., Mirabedini, S. J., & Abtahi, A. (2018). Designing a Model for Implementation of Business Intelligence in the Banking Industry. *International Journal of Enterprise Information Systems*, 14(1), 77–103. doi:10.4018/IJEIS.2018010105

Rezende, D. A. (2018). Strategic Digital City Projects: Innovative Information and Public Services Offered by Chicago (USA) and Curitiba (Brazil). In M. Lytras, L. Daniela, & A. Visvizi (Eds.), *Enhancing Knowledge Discovery and Innovation in the Digital Era* (pp. 204–223). Hershey, PA: IGI Global. doi:10.4018/978-1-5225-4191-2.ch012

Rodriguez, A., Rico-Diaz, A. J., Rabuñal, J. R., & Gestal, M. (2017). Fish Tracking with Computer Vision Techniques: An Application to Vertical Slot Fishways. In M. S., & V. V. (Eds.), Multi-Core Computer Vision and Image Processing for Intelligent Applications (pp. 74-104). Hershey, PA: IGI Global. https://doi.org/doi:10.4018/978-1-5225-0889-2.ch003

Romero, J. A. (2018). Sustainable Advantages of Business Value of Information Technology. In M. Khosrow-Pour, D.B.A. (Ed.), Encyclopedia of Information Science and Technology, Fourth Edition (pp. 923-929). Hershey, PA: IGI Global. doi:10.4018/978-1-5225-2255-3.ch079

Romero, J. A. (2018). The Always-On Business Model and Competitive Advantage. In N. Bajgoric (Ed.), *Always-On Enterprise Information Systems for Modern Organizations* (pp. 23–40). Hershey, PA: IGI Global. doi:10.4018/978-1-5225-3704-5.ch002

Rosen, Y. (2018). Computer Agent Technologies in Collaborative Learning and Assessment. In M. Khosrow-Pour, D.B.A. (Ed.), Encyclopedia of Information Science and Technology, Fourth Edition (pp. 2402-2410). Hershey, PA: IGI Global. doi:10.4018/978-1-5225-2255-3.ch209

Roy, D. (2018). Success Factors of Adoption of Mobile Applications in Rural India: Effect of Service Characteristics on Conceptual Model. In M. Khosrow-Pour (Ed.), *Green Computing Strategies for Competitive Advantage and Business Sustainability* (pp. 211–238). Hershey, PA: IGI Global. doi:10.4018/978-1-5225-5017-4.ch010

Ruffin, T. R., & Hawkins, D. P. (2018). Trends in Health Care Information Technology and Informatics. In M. Khosrow-Pour, D.B.A. (Ed.), Encyclopedia of Information Science and Technology, Fourth Edition (pp. 3805-3815). Hershey, PA: IGI Global. doi:10.4018/978-1-5225-2255-3.ch330

Sadasivam, U. M., & Ganesan, N. (2021). Detecting Fake News Using Deep Learning and NLP. In S. Misra, C. Arumugam, S. Jaganathan, & S. S. (Eds.), *Confluence of AI, Machine, and Deep Learning in Cyber Forensics* (pp. 117-133). IGI Global. https://doi.org/10.4018/978-1-7998-4900-1.ch007

Safari, M. R., & Jiang, Q. (2018). The Theory and Practice of IT Governance Maturity and Strategies Alignment: Evidence From Banking Industry. *Journal of Global Information Management*, 26(2), 127–146. doi:10.4018/JGIM.2018040106

Sahin, H. B., & Anagun, S. S. (2018). Educational Computer Games in Math Teaching: A Learning Culture. In E. Toprak & E. Kumtepe (Eds.), *Supporting Multiculturalism in Open and Distance Learning Spaces* (pp. 249–280). Hershey, PA: IGI Global. doi:10.4018/978-1-5225-3076-3.ch013

Sakalle, A., Tomar, P., Bhardwaj, H., & Sharma, U. (2021). Impact and Latest Trends of Intelligent Learning With Artificial Intelligence. In S. Verma & P. Tomar (Eds.), *Impact of AI Technologies on Teaching, Learning, and Research in Higher Education* (pp. 172-189). IGI Global. https://doi.org/10.4018/978-1-7998-4763-2.ch011

Sala, N. (2021). Virtual Reality, Augmented Reality, and Mixed Reality in Education: A Brief Overview. In D. Choi, A. Dailey-Hebert, & J. Estes (Eds.), *Current and Prospective Applications of Virtual Reality in Higher Education* (pp. 48–73). IGI Global. https://doi.org/10.4018/978-1-7998-4960-5.ch003

Salunkhe, S., Kanagachidambaresan, G., Rajkumar, C., & Jayanthi, K. (2022). Online Detection and Prediction of Fused Deposition Modelled Parts Using Artificial Intelligence. In S. Salunkhe, H. Hussein, & J. Davim (Eds.), *Applications of Artificial Intelligence in Additive Manufacturing* (pp. 194–209). IGI Global. https://doi.org/10.4018/978-1-7998-8516-0.ch009

Samy, V. S., Pramanick, K., Thenkanidiyoor, V., & Victor, J. (2021). Data Analysis and Visualization in Python for Polar Meteorological Data. *International Journal of Data Analytics*, 2(1), 32–60. https://doi.org/10.4018/IJDA.2021010102

Sanna, A., & Valpreda, F. (2017). An Assessment of the Impact of a Collaborative Didactic Approach and Students' Background in Teaching Computer Animation. *International Journal of Information and Communication Technology Education*, 13(4), 1–16. doi:10.4018/IJICTE.2017100101

Sarivougioukas, J., & Vagelatos, A. (2022). Fused Contextual Data With Threading Technology to Accelerate Processing in Home UbiHealth. *International Journal of Software Science and Computational Intelligence*, 14(1), 1–14. https://doi.org/10.4018/IJSSCI.285590

Scott, A., Martin, A., & McAlear, F. (2017). Enhancing Participation in Computer Science among Girls of Color: An Examination of a Preparatory AP Computer Science Intervention. In Y. Rankin & J. Thomas (Eds.), *Moving Students of Color from Consumers to Producers of Technology* (pp. 62–84). Hershey, PA: IGI Global. doi:10.4018/978-1-5225-2005-4.ch004

Shanmugam, M., Ibrahim, N., Gorment, N. Z., Sugu, R., Dandarawi, T. N., & Ahmad, N. A. (2022). Towards an Integrated Omni-Channel Strategy Framework for Improved Customer Interaction. In P. Lai (Ed.), *Handbook of Research on Social Impacts of E-Payment and Blockchain Technology* (pp. 409–427). IGI Global. https://doi.org/10.4018/978-1-7998-9035-5.ch022

Sharma, A., & Kumar, S. (2021). Network Slicing and the Role of 5G in IoT Applications. In S. Kumar, M. Trivedi, P. Ranjan, & A. Punhani (Eds.), *Evolution of Software-Defined Networking Foundations for IoT and 5G Mobile Networks* (pp. 172–190). IGI Global. https://doi.org/10.4018/978-1-7998-4685-7.ch010

Siddoo, V., & Wongsai, N. (2017). Factors Influencing the Adoption of ISO/IEC 29110 in Thai Government Projects: A Case Study. *International Journal of Information Technologies and Systems Approach, 10*(1), 22–44. doi:10.4018/IJITSA.2017010102

Silveira, C., Hir, M. E., & Chaves, H. K. (2022). An Approach to Information Management as a Subsidy of Global Health Actions: A Case Study of Big Data in Health for Dengue, Zika, and Chikungunya. In J. Lima de Magalhães, Z. Hartz, G. Jamil, H. Silveira, & L. Jamil (Eds.), *Handbook of Research on Essential Information Approaches to Aiding Global Health in the One Health Context* (pp. 219–234). IGI Global. https://doi.org/10.4018/978-1-7998-8011-0.ch012

Simões, A. (2017). Using Game Frameworks to Teach Computer Programming. In R. Alexandre Peixoto de Queirós & M. Pinto (Eds.), *Gamification-Based E-Learning Strategies for Computer Programming Education* (pp. 221–236). Hershey, PA: IGI Global. doi:10.4018/978-1-5225-1034-5.ch010

Simões de Almeida, R., & da Silva, T. (2022). AI Chatbots in Mental Health: Are We There Yet? In A. Marques & R. Queirós (Eds.), *Digital Therapies in Psychosocial Rehabilitation and Mental Health* (pp. 226–243). IGI Global. https://doi.org/10.4018/978-1-7998-8634-1.ch011

Singh, L. K., Khanna, M., Thawkar, S., & Gopal, J. (2021). Robustness for Authentication of the Human Using Face, Ear, and Gait Multimodal Biometric System. *International Journal of Information System Modeling and Design, 12*(1), 39–72. https://doi.org/10.4018/IJISMD.2021010103

Sllame, A. M. (2017). Integrating LAB Work With Classes in Computer Network Courses. In H. Alphin Jr, R. Chan, & J. Lavine (Eds.), *The Future of Accessibility in International Higher Education* (pp. 253–275). Hershey, PA: IGI Global. doi:10.4018/978-1-5225-2560-8.ch015

Smirnov, A., Ponomarev, A., Shilov, N., Kashevnik, A., & Teslya, N. (2018). Ontology-Based Human-Computer Cloud for Decision Support: Architecture and Applications in Tourism. *International Journal of Embedded and Real-Time Communication Systems, 9*(1), 1–19. doi:10.4018/IJERTCS.2018010101

Smith-Ditizio, A. A., & Smith, A. D. (2018). Computer Fraud Challenges and Its Legal Implications. In M. Khosrow-Pour, D.B.A. (Ed.), Encyclopedia of Information Science and Technology, Fourth Edition (pp. 4837-4848). Hershey, PA: IGI Global. doi:10.4018/978-1-5225-2255-3.ch419

Sosnin, P. (2018). Figuratively Semantic Support of Human-Computer Interactions. In *Experience-Based Human-Computer Interactions: Emerging Research and Opportunities* (pp. 244–272). Hershey, PA: IGI Global. doi:10.4018/978-1-5225-2987-3.ch008

Srilakshmi, R., & Jaya Bhaskar, M. (2021). An Adaptable Secure Scheme in Mobile Ad hoc Network to Protect the Communication Channel From Malicious Behaviours. *International Journal of Information Technology and Web Engineering, 16*(3), 54–73. https://doi.org/10.4018/IJITWE.2021070104

Sukhwani, N., Kagita, V. R., Kumar, V., & Panda, S. K. (2021). Efficient Computation of Top-K Skyline Objects in Data Set With Uncertain Preferences. *International Journal of Data Warehousing and Mining, 17*(3), 68–80. https://doi.org/10.4018/IJDWM.2021070104

Susanto, H., Yie, L. F., Setiana, D., Asih, Y., Yoganingrum, A., Riyanto, S., & Saputra, F. A. (2021). Digital Ecosystem Security Issues for Organizations and Governments: Digital Ethics and Privacy. In Z. Mahmood (Ed.), *Web 2.0 and Cloud Technologies for Implementing Connected Government* (pp. 204–228). IGI Global. https://doi.org/10.4018/978-1-7998-4570-6.ch010

Syväjärvi, A., Leinonen, J., Kivivirta, V., & Kesti, M. (2017). The Latitude of Information Management in Local Government: Views of Local Government Managers. *International Journal of Electronic Government Research, 13*(1), 69–85. doi:10.4018/IJEGR.2017010105

Tanque, M., & Foxwell, H. J. (2018). Big Data and Cloud Computing: A Review of Supply Chain Capabilities and Challenges. In A. Prasad (Ed.), *Exploring the Convergence of Big Data and the Internet of Things* (pp. 1–28). Hershey, PA: IGI Global. doi:10.4018/978-1-5225-2947-7.ch001

Teixeira, A., Gomes, A., & Orvalho, J. G. (2017). Auditory Feedback in a Computer Game for Blind People. In T. Issa, P. Kommers, T. Issa, P. Isaías, & T. Issa (Eds.), *Smart Technology Applications in Business Environments* (pp. 134–158). Hershey, PA: IGI Global. doi:10.4018/978-1-5225-2492-2.ch007

Tewari, P., Tiwari, P., & Goel, R. (2022). Information Technology in Supply Chain Management. In V. Garg & R. Goel (Eds.), *Handbook of Research on Innovative Management Using AI in Industry 5.0* (pp. 165–178). IGI Global. https://doi.org/10.4018/978-1-7998-8497-2.ch011

Thompson, N., McGill, T., & Murray, D. (2018). Affect-Sensitive Computer Systems. In M. Khosrow-Pour, D.B.A. (Ed.), Encyclopedia of Information Science and Technology, Fourth Edition (pp. 4124-4135). Hershey, PA: IGI Global. doi:10.4018/978-1-5225-2255-3.ch357

Triberti, S., Brivio, E., & Galimberti, C. (2018). On Social Presence: Theories, Methodologies, and Guidelines for the Innovative Contexts of Computer-Mediated Learning. In M. Marmon (Ed.), *Enhancing Social Presence in Online Learning Environments* (pp. 20–41). Hershey, PA: IGI Global. doi:10.4018/978-1-5225-3229-3.ch002

Tripathy, B. K. T. R., S., & Mohanty, R. K. (2018). Memetic Algorithms and Their Applications in Computer Science. In S. Dash, B. Tripathy, & A. Rahman (Eds.), Handbook of Research on Modeling, Analysis, and Application of Nature-Inspired Metaheuristic Algorithms (pp. 73-93). Hershey, PA: IGI Global. https://doi.org/doi:10.4018/978-1-5225-2857-9.ch004

Turulja, L., & Bajgoric, N. (2017). Human Resource Management IT and Global Economy Perspective: Global Human Resource Information Systems. In M. Khosrow-Pour (Ed.), *Handbook of Research on Technology Adoption, Social Policy, and Global Integration* (pp. 377–394). Hershey, PA: IGI Global. doi:10.4018/978-1-5225-2668-1.ch018

Unwin, D. W., Sanzogni, L., & Sandhu, K. (2017). Developing and Measuring the Business Case for Health Information Technology. In K. Moahi, K. Bwalya, & P. Sebina (Eds.), *Health Information Systems and the Advancement of Medical Practice in Developing Countries* (pp. 262–290). Hershey, PA: IGI Global. doi:10.4018/978-1-5225-2262-1.ch015

Usharani, B. (2022). House Plant Leaf Disease Detection and Classification Using Machine Learning. In M. Mundada, S. Seema, S. K.G., & M. Shilpa (Eds.), *Deep Learning Applications for Cyber-Physical Systems* (pp. 17-26). IGI Global. https://doi.org/10.4018/978-1-7998-8161-2.ch002

Vadhanam, B. R. S., M., Sugumaran, V., V., V., & Ramalingam, V. V. (2017). Computer Vision Based Classification on Commercial Videos. In M. S., & V. V. (Eds.), Multi-Core Computer Vision and Image Processing for Intelligent Applications (pp. 105-135). Hershey, PA: IGI Global. https://doi.org/doi:10.4018/978-1-5225-0889-2.ch004

Vairinho, S. (2022). Innovation Dynamics Through the Encouragement of Knowledge Spin-Off From Touristic Destinations. In C. Ramos, S. Quinteiro, & A. Gonçalves (Eds.), *ICT as Innovator Between Tourism and Culture* (pp. 170–190). IGI Global. https://doi.org/10.4018/978-1-7998-8165-0.ch011

Valverde, R., Torres, B., & Motaghi, H. (2018). A Quantum NeuroIS Data Analytics Architecture for the Usability Evaluation of Learning Management Systems. In S. Bhattacharyya (Ed.), *Quantum-Inspired Intelligent Systems for Multimedia Data Analysis* (pp. 277–299). Hershey, PA: IGI Global. doi:10.4018/978-1-5225-5219-2.ch009

Vassilis, E. (2018). Learning and Teaching Methodology: "1:1 Educational Computing. In K. Koutsopoulos, K. Doukas, & Y. Kotsanis (Eds.), *Handbook of Research on Educational Design and Cloud Computing in Modern Classroom Settings* (pp. 122–155). Hershey, PA: IGI Global. doi:10.4018/978-1-5225-3053-4.ch007

Verma, S., & Jain, A. K. (2022). A Survey on Sentiment Analysis Techniques for Twitter. In B. Gupta, D. Peraković, A. Abd El-Latif, & D. Gupta (Eds.), *Data Mining Approaches for Big Data and Sentiment Analysis in Social Media* (pp. 57–90). IGI Global. https://doi.org/10.4018/978-1-7998-8413-2.ch003

Wang, H., Huang, P., & Chen, X. (2021). Research and Application of a Multidimensional Association Rules Mining Method Based on OLAP. *International Journal of Information Technology and Web Engineering, 16*(1), 75–94. https://doi.org/10.4018/IJITWE.2021010104

Wexler, B. E. (2017). Computer-Presented and Physical Brain-Training Exercises for School Children: Improving Executive Functions and Learning. In B. Dubbels (Ed.), *Transforming Gaming and Computer Simulation Technologies across Industries* (pp. 206–224). Hershey, PA: IGI Global. doi:10.4018/978-1-5225-1817-4.ch012

Wimble, M., Singh, H., & Phillips, B. (2018). Understanding Cross-Level Interactions of Firm-Level Information Technology and Industry Environment: A Multilevel Model of Business Value. *Information Resources Management Journal, 31*(1), 1–20. doi:10.4018/IRMJ.2018010101

Wimmer, H., Powell, L., Kilgus, L., & Force, C. (2017). Improving Course Assessment via Web-based Homework. *International Journal of Online Pedagogy and Course Design, 7*(2), 1–19. doi:10.4018/IJOPCD.2017040101

Wong, S. (2021). Gendering Information and Communication Technologies in Climate Change. In M. Khosrow-Pour D.B.A. (Eds.), *Encyclopedia of Information Science and Technology, Fifth Edition* (pp. 1408-1422). IGI Global. https://doi.org/10.4018/978-1-7998-3479-3.ch096

Wong, Y. L., & Siu, K. W. (2018). Assessing Computer-Aided Design Skills. In M. Khosrow-Pour, D.B.A. (Ed.), Encyclopedia of Information Science and Technology, Fourth Edition (pp. 7382-7391). Hershey, PA: IGI Global. doi:10.4018/978-1-5225-2255-3.ch642

Wongsurawat, W., & Shrestha, V. (2018). Information Technology, Globalization, and Local Conditions: Implications for Entrepreneurs in Southeast Asia. In P. Ordóñez de Pablos (Ed.), *Management Strategies and Technology Fluidity in the Asian Business Sector* (pp. 163–176). Hershey, PA: IGI Global. doi:10.4018/978-1-5225-4056-4.ch010

Yamada, H. (2021). Homogenization of Japanese Industrial Technology From the Perspective of R&D Expenses. *International Journal of Systems and Service-Oriented Engineering, 11*(2), 24–51. doi:10.4018/IJSSOE.2021070102

Yang, Y., Zhu, X., Jin, C., & Li, J. J. (2018). Reforming Classroom Education Through a QQ Group: A Pilot Experiment at a Primary School in Shanghai. In H. Spires (Ed.), *Digital Transformation and Innovation in Chinese Education* (pp. 211–231). Hershey, PA: IGI Global. doi:10.4018/978-1-5225-2924-8.ch012

Yilmaz, R., Sezgin, A., Kurnaz, S., & Arslan, Y. Z. (2018). Object-Oriented Programming in Computer Science. In M. Khosrow-Pour, D.B.A. (Ed.), Encyclopedia of Information Science and Technology, Fourth Edition (pp. 7470-7480). Hershey, PA: IGI Global. doi:10.4018/978-1-5225-2255-3.ch650

Yu, L. (2018). From Teaching Software Engineering Locally and Globally to Devising an Internationalized Computer Science Curriculum. In S. Dikli, B. Etheridge, & R. Rawls (Eds.), *Curriculum Internationalization and the Future of Education* (pp. 293–320). Hershey, PA: IGI Global. doi:10.4018/978-1-5225-2791-6.ch016

Yuhua, F. (2018). Computer Information Library Clusters. In M. Khosrow-Pour, D.B.A. (Ed.), Encyclopedia of Information Science and Technology, Fourth Edition (pp. 4399-4403). Hershey, PA: IGI Global. doi:10.4018/978-1-5225-2255-3.ch382

Zakaria, R. B., Zainuddin, M. N., & Mohamad, A. H. (2022). Distilling Blockchain: Complexity, Barriers, and Opportunities. In P. Lai (Ed.), *Handbook of Research on Social Impacts of E-Payment and Blockchain Technology* (pp. 89–114). IGI Global. https://doi.org/10.4018/978-1-7998-9035-5.ch007

Zhang, Z., Ma, J., & Cui, X. (2021). Genetic Algorithm With Three-Dimensional Population Dominance Strategy for University Course Timetabling Problem. *International Journal of Grid and High Performance Computing, 13*(2), 56–69. https://doi.org/10.4018/IJGHPC.2021040104

About the Contributors

Eduard Babulak is accomplished international scholar, researcher, consultant, educator, professional engineer and polyglot, with more than thirty years of experience. He served as successfully published and his research was cited by scholars all over the world. He speaks 16 languages and his biography was cited in the Cambridge Blue Book, Cambridge Index of Biographies, Stanford Who's Who, and number of issues of Who's Who in the World and America.

* * *

C. V. Suresh Babu is a pioneer in content development. A true entrepreneur, he founded Anniyappa Publications, a company that is highly active in publishing books related to Computer Science and Management. Dr. C.V. Suresh Babu has also ventured into SB Institute, a center for knowledge transfer. He holds a Ph.D. in Engineering Education from the National Institute of Technical Teachers Training & Research in Chennai, along with seven master's degrees in various disciplines such as Engineering, Computer Applications, Management, Commerce, Economics, Psychology, Law, and Education. Additionally, he has UGC-NET/SET qualifications in the fields of Computer Science, Management, Commerce, and Education. Currently, Dr. C.V. Suresh Babu is a Professor in the Department of Information Technology at the School of Computing Science, Hindustan Institute of Technology and Science (Hindustan University) in Padur, Chennai, Tamil Nadu, India. For more information, you can visit his personal blog at .

M. Beulah Viji Christiana, Professor in the Department of Master of Business Administration, Panimalar Engineering College affiliated to Anna University, has a got a rich teaching experience which spans over 24 years. She is an experienced and innovative student focused professional and mentor with a proven track record of excellent teaching skills. Her academic qualification encompasses Post Graduate degrees in English Literature and Business Administration from Madurai Kamaraj University and she received her M.Phil in Management Studies from Bharathidasan

University. She was awarded Ph.D in Management Studies from Mother Teresa Women's University, Kodaikanal in 2012. She has published more than 40 research articles in various national and international journals of repute as well as in conferences. Known for her extensive exposure in mentoring students through igniting their creative and innovative ability as well in stimulating them to think logically and analytically, she has guided many students in their research projects pertaining to the various functional areas of management. Besides she is also known for her role as a reviewer in various journals and books and she has authored five books .

Sampath Boopathi is an accomplished individual with a strong academic background and extensive research experience. He completed his undergraduate studies in Mechanical Engineering and pursued his postgraduate studies in the field of Computer-Aided Design. Dr. Boopathi obtained his Ph.D. from Anna University, focusing his research on Manufacturing and optimization. Throughout his career, Dr. Boopathi has made significant contributions to the field of engineering. He has authored and published over 200 research articles in internationally peer-reviewed journals, highlighting his expertise and dedication to advancing knowledge in his area of specialization. His research output demonstrates his commitment to conducting rigorous and impactful research. In addition to his research publications, Dr. Boopathi has also been granted one patent and has three published patents to his name. This indicates his innovative thinking and ability to develop practical solutions to real-world engineering challenges. With 17 years of academic and research experience, Dr. Boopathi has enriched the engineering community through his teaching and mentorship roles.

B. Charith has 19+ years of total work experience and specifically 8+ yrs. in the leadership hiring with proven results and repeated success in multiple units and assignment handled, building good client relationship and value, driving vision and achieving critical strategic goals towards the Organization's growth. Valued contributor to key strategic improvements and highly successful new set ups. Personally hold a Track record in placing Leaders for leading Giants in India across functions like ITES operations, HR, Quality, Transition, Presales and Solution Design etc. Been instrumental in closing Niche mandates at Leadership band and successful on boarding of critical resources who are now in the Executive Band of Fortune 100 Companies. Apart from Heading Leadership hiring directly, personally being a part of the Golden Opportunities Management group, being instrumental in terms of important decision making, Process Improvement, MIS, Overall key account management and Global Revenue management as well.

Sahil Manoj Chavan is a faculty member in the Department of Electrical Power System at Sandip University, located in Nashik, Maharashtra, India.

Yagya Dutta Dwivedi is a faculty member in the Department of Aeronautical Engineering at the Institute of Aeronautical Engineering, located in Hyderabad, Telangana, India.

Sureshkumar Myilsamy completed his undergraduate in Mechanical Engineering and postgraduate in the field of Engineering Design. He completed his Ph.D. from Anna University, Chennai, Tamil Nādu, India.

Hema N. is currently working as Associate Professor in Dept. of ISE, RNSIT. She is having 15 years of Teaching and 8 years of Research experience. She has guided various under graduate and MTech projects. She has involved in various Academic co ordinations. She has pursued Ph.D from VTU in the area of Medical Image Processing. She has published more than 15 papers in various national and international conferences and Journals. She has published one book chapter and one patent.

Narayanaswamy Venkateswaran works as Professor at Panimalar Engineering College, having 26 years of teaching experience published more papers in International journals and also member of editorial board in more reputed journals. Published books on International Business Management, HR Analytics, Entrepreneurship Development, Professional Ethics and Human Values, Fundamentals of Management. Area of expertise is on Business Analytics, Artificial Intelligence and Supply Chain Management.

Index

A

AI 1-8, 11-18, 22-28, 30-35, 37-61, 63-67, 69-83, 86-87, 89-90, 92-93, 114-116, 118, 120-123, 126-132, 134-144, 171-173, 175-177, 179, 182-184, 186, 188, 193, 195-205, 207-226, 228, 230
AI in Education 63
AI Integration 74, 120, 122-123, 196, 205
AI-Powered Humanoid Robots 4, 11, 14, 16, 22-23, 31-32, 85-90, 92-95, 98-102, 104, 106-110, 112-115
AM 144
Applied AI 1-5, 32, 37-38, 57, 59, 200-204, 210-211, 217, 224-225
AR 28, 64, 135, 144, 174-175, 195, 207, 216, 221
Artificial Intelligence 1-2, 5, 7-8, 12, 14, 24, 27-28, 30-35, 37-38, 59-62, 70, 78, 82-84, 90, 100, 105, 115-119, 121, 123, 127-129, 138, 142, 144, 170-171, 173, 175, 189, 194, 198-200, 203, 207, 215, 227-231
Augmented Reality 28, 64, 135, 144, 170, 174, 207, 221
Automation 12-14, 26, 31-33, 36, 40-42, 45, 48, 57, 59-60, 63-64, 69, 80, 82, 90, 98, 118, 121-122, 126-128, 138, 140, 171-173, 182, 186, 193, 195, 202, 213, 227

B

Blockchain in Education 63, 74-75
Business 34, 37-39, 41-42, 44, 50, 56, 58-62, 67, 72, 76-77, 80, 83, 115, 117, 120, 146, 168-170, 172, 179, 184-186, 189-191, 193-194, 202, 228

C

Challenges 1, 4, 6, 10, 22-23, 31-32, 34, 38-39, 54-56, 59, 61, 66-75, 77, 80, 86, 88-89, 93, 112, 118-119, 121-123, 128-129, 135, 137, 141, 143-144, 170-172, 174-175, 179, 183, 186-188, 193-198, 200-201, 212, 215, 222-230
Chat Bots 34
Chatbots 8, 13, 18, 23, 34, 38-39, 41, 43, 48-49, 57, 173, 185, 202
Cognitive Capabilities 86-87, 91-93
Curriculum Integration 63
Customer Experience 48, 185, 228
Cybersecurity 28, 57, 80, 83, 118-119, 121, 171, 175-176, 186, 189, 193, 197, 212, 214, 220-221, 228
Cyberspace 170, 172, 195, 201, 204, 207-209, 211-212, 217, 219, 224-225, 228

E

Education 4, 11, 15-18, 22, 24, 32-34, 63-65, 67, 71-76, 78-80, 82, 84, 86-87, 89-90, 92, 103, 107, 113, 115, 118-120, 172, 174, 193-194, 197-198, 200, 202, 204, 207, 209, 211-213, 216, 224-225, 228
Efficiency 1, 4, 8, 11-12, 14, 23, 25, 27, 34, 37-41, 44-45, 49, 54, 59, 65, 67, 74-76, 80, 86, 88, 90-92, 96-97, 111, 114-115, 121-123, 125-129, 131-132,

134, 136-139, 141, 147, 170, 173, 180, 182-185, 189-190, 193, 196, 203, 211, 213, 217-218, 221, 224, 228

Emotional Intelligence 86-87, 89, 99-100, 114-115, 117, 216

Employment Patterns 63, 69

Entertainment 1-2, 11, 14, 32, 34-36, 86-87, 89-90, 107, 174, 200, 202, 205, 212-213, 219

Ethical Considerations 1, 4, 11, 31-32, 37-38, 50, 53-54, 56, 59, 63, 66, 71-72, 77, 81, 86, 112, 115, 140-141, 171, 179, 191, 193-194, 216

Ethical Implications 73, 121, 123, 179

Ethics 50, 63, 76, 80, 82-83, 117, 179-180, 186, 193

EV 136, 144

F

Fatigue 122, 125, 136, 146-149, 151, 154, 156, 167-168

Freezer Drawer 145-146, 160-165, 167

Friction Stir Welding 120-124, 128-129, 132, 136-144

FSW 120-125, 128-141, 143-144

Future Directions 4, 9, 61, 86, 88-89, 113, 123, 137, 141, 225

Future Trends 4, 37, 54-55, 57, 59-60, 117

FWS 141, 144

G

Generative AI 35, 118, 198

GUI 144, 229

H

Healthcare 1, 4, 7-12, 15, 22, 25, 28, 32-39, 41, 53, 60-62, 80, 86-87, 89-90, 92, 96, 103, 107, 113, 115-116, 119, 171-174, 182-183, 193, 195-197, 200-202, 204-205, 209-212, 215-216, 219, 223-225, 227-230

Humanoid Robotics 1-5, 7, 11-12, 14, 16-17, 22-26, 31-33, 200-205, 212, 224-225

Humanoid Robots 1-8, 11, 14-17, 22-26, 31-33, 85-104, 106-110, 112-115, 118, 120-123, 129-136, 140-141, 202, 225, 230

Human-Robot Interaction 1, 3, 23-24, 26, 31-33, 86-91, 101-102, 104-108, 112, 114-115, 117, 126, 135, 138, 204, 206-207, 214, 225-227, 229

I

IMU 144

Industrial Robotics 33, 63, 80, 83

Industry 4, 14, 31-33, 37-39, 50, 60-62, 69-70, 72-73, 76-77, 79, 83, 90, 118-119, 122, 124-126, 136, 141, 143, 171, 174, 179, 182-183, 185-186, 188-189, 191, 197-198, 228

Industry-Academia Collaboration 76

Innovation 4, 12, 23, 25-26, 32, 35, 37-39, 53-54, 58-59, 64-65, 71-73, 76, 82, 84-85, 87-88, 90, 112, 121-123, 125, 127-128, 136-137, 171-180, 186-194, 199, 211, 213, 224

M

Machine Learning 2, 6, 14, 27, 34, 38, 43-44, 47, 56-57, 61-62, 66, 70, 86-87, 93, 103, 105-110, 112, 115-116, 118-119, 126-127, 129, 131, 134, 138-139, 171, 173, 177-179, 183, 185-186, 188-189, 194, 199, 204, 206, 216-217, 223, 230

Manufacturing 1, 4, 11-14, 25, 32-33, 35-40, 80, 86-87, 90, 92, 96-97, 115, 121-127, 130-132, 135-138, 140-141, 143-144, 182, 195-196, 199-200, 202, 205, 212, 223

Mechanical Product 145, 149-150, 167

Motor Skills 86-88, 94-96, 98

N

Networking Technologies 170

O

Optimization 28, 37-40, 43, 47, 49, 60-62, 107, 111-113, 115, 118, 126, 128-129, 131, 134, 136, 138-139, 142-143, 171, 173-174, 178, 184-185, 191-192, 195

P

Parametric ALT 145-146, 148-149, 151, 165
Performance Enhancement 85, 88-89
Personalization 7, 11, 41, 46-47, 59, 63, 66, 82, 108, 116, 170, 184, 208, 217
Predictive Maintenance 27, 38-40, 121, 131-132, 138, 202, 213
Privacy 4, 22, 31, 38, 51-56, 58-59, 64, 66-67, 69, 71-74, 81, 86, 112-115, 119, 140-141, 171, 175, 179-180, 186, 189, 191, 193, 198, 203, 211, 214, 218, 222, 225

R

Robotics 1-5, 7, 11-12, 14, 16-17, 22-26, 31-35, 61, 63, 80, 83, 86-87, 90, 92, 96, 114-118, 121-123, 125-129, 134, 137-138, 141-143, 195-198, 200-205, 207, 209-210, 212-214, 224-226, 228-231

S

Sample Size 145, 148-149, 154, 157-159, 167, 169

T

Transparency 4, 30, 50-53, 56, 58-59, 63, 66-69, 74-75, 78-79, 81-82, 107, 140-141, 179, 186, 191, 193-194, 212, 214, 225
Trust 31, 50-54, 56, 58-59, 63, 66-68, 71, 74-75, 79, 81, 83, 87, 91, 103-104, 112-113, 119, 175, 179, 183, 186, 198, 208, 214-215, 223, 229-230

U

Ultra-Smart Computing 170-174, 176-194
Ultra-Smart Cyberspace 200, 207-209, 217, 224-225

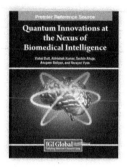

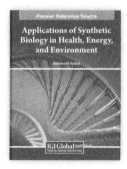

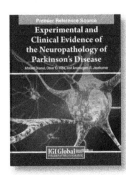
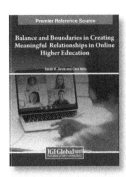

Printed in the USA
CPSIA information can be obtained
at www.ICGtesting.com
LVHW080533170924
791295LV00005B/406

9 798369 323991